INSIGHT INTO CALCULUS USING TEXAS INSTRUMENTS GRAPHICS CALCULATORS

FRANK WARD
INDIAN RIVER COMMUNITY COLLEGE

DOUG WILBERSCHEID
INDIAN RIVER COMMUNITY COLLEGE

PRENTICE HALL Upper Saddle River, NJ 07458

Acquisitions Editor: *George Lobell*
Production Editor: *Lorena Cerisano*
Supplement Editor: *Audra Walsh*
Special Projects Manager: *Barbara A. Murray*
Production Coordinator: *Alan Fischer*

Printed in the United States of America

10 9 8 7 6 5 4 3 2 1

ISBN 0-13-255365-1

Prentice-Hall International (UK) Limited, *London*
Prentice-Hall of Australia Pty. Limited, *Sydney*
Prentice-Hall Canada, Inc., *Toronto*
Prentice-Hall Hispanoamericana, S.A., *Mexico*
Prentice-Hall of India Private Limited, *New Delhi*
Prentice-Hall of Japan, Inc., *Tokyo*
Simon & Schuster Asia Pte. Ltd., *Singapore*
Editora Prentice-Hall do Brasil, Ltda., *Rio de Janeiro*

Preface

The main purpose of the projects in this book is to help you gain a better understanding of calculus through a guided discovery approach Discovery requires a high level of active learning. We believe that the discovery projects in this book contain enough clues to assure a successful conclusion, if you are willing to actively participate in your own mathematical development.

The graphing calculator does not eliminate the need for a solid understanding of calculus concepts, but is instead used here to provide insight into those concepts. Graphing calculators allow us to solve problems using both numerical and graphical methods, and also to use these methods to reinforce the analytical techniques which you will learn in class. A graphing calculator is a powerful tool if used properly. If not used properly, it can lead to false conclusions and may be a stumbling block to understanding a concept or technique.

Curve fitting is a method by which we find a mathematical function to describe a set of data which corresponds to a real-life situation. Until recent years, the link between these mathematical functions and the real-life situation was lost due to the complexity of the curve-fitting process. Curve-fitting was done by scientists and mathematicians, who presented the final product to students in the form of functions. With the advent of graphing calculators, this link between functions and the real situations they represent can be easily made by you, the student.

Using this link, many of the projects in this book involve the application of calculus concepts to real data, which makes the projects more interesting and the concepts more pertinent. There are several projects which require you to collect data on the Texas Instruments Calculator Based Laboratory (CBL).

It is our hope that through these projects you will gain confidence in your problem solving skills and in the use of technology in appropriate situations. The projects in this book require you to take an active role in the learning process. An active learner reads about mathematics, does mathematics, and talks about mathematics. We recommend that you:

1. Work through each given example in detail and check your results with ours. Clarity and insight come from working through each example step by step.

2. Use your observation power to look for graphical, numerical, and algebraic patterns.

3. Form a connection between the project and the calculus concepts which you are learning in class.

4. Take a few minutes at the end of the project to

think about your results and, if possible, discuss your results with a classmate. Many ideas and insights are the result of an active two-way discussion. Little things that are said sometimes trigger a thought or an idea that leads to a deeper understanding of a concept.

The projects in this book are not designed to teach you how to use your calculator. When a project requires you to perform a complicated calculator operation, you will be referred to the appendix, where the specific keystrokes are described in detail for TI-82, TI-85, and TI-92 calculators.

The projects of this book are designed to help you develop the skills of thinking analytically, recognizing patterns, forming conclusions, and organizing your results in a clear manner. Corporations look for potential employees who possess these characteristics, so take advantage of this opportunity to develop and sharpen these skills.

Supplementary Materials

- **Instructor's Guide** provides expected learning outcomes, logistics, and guidelines for acceptable solutions for every project, as well as CBL troubleshooting tips for the CBL projects. The instructor's guide also contains an applications index.
- **Internet access to programs** for the TI-82, TI-85, and TI-92 graphing calculators which are designed for particular projects. These programs can be found in the World Wide Web at http://www.ircc.cc.fl.us.
- **Program Disk for Instructors** contains the same information found on the internet, for those who do not have access to the internet.

Acknowledgments

We are grateful for the support of the National Science Foundation and to the faculty and administration of Indian River Community College. The projects of this book have benefitted greatly from the dedicated work of the reviewers, programmers, typists, and artists who have also been premier math students at Indian River Community College: Sophia Wilberscheid, Zina Smith, David Donner, John Hunter, Chris Wells, Joshua Wirth, and José Garcia.

Precalculus

Limits

Derivatives

Integrals

Polar Coordinates

Sequences and Series

Curve Fitting

Appendices

Objectives:
1. Use vertical and horizontal translations to fit a function
    to a set of data.
2. Use piecewise-defined functions to model a set of data.

Technology:
    TI-82, TI-85 (CBL compatable version), or TI-92 graphing
    calculator, CBL, CBL motion detector, and program DROPX.

Prerequisites:
    Introduction to vertical and horizontal translations and
    piecewise-defined functions.

Overview: If we drop a ball from a height of exactly 8 feet at
a time of exactly x = 0 seconds, the height of the ball, in feet,
can be described by $y = -16x^2 + 8$, where x is time in seconds.
We are assuming that air resistance is negligible. If we drop
the ball from a height which is different than 8 feet, and at a
time which is different than x = 0 seconds, we can use horizontal
and vertical translations to adjust the model. We will attempt
to model the motion of the ball as it falls, bounces up, and
falls again (see Figure 1).

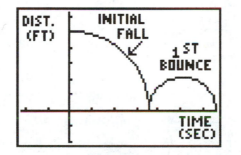

Figure 1

We will fit a function to each of the arcs separately, and then
"paste" these two functions together in the form of a piecewise-
defined function.

Procedures:
1. Attach the motion detector to the ceiling as shown in Figure
    2, with the beam facing directly downward.

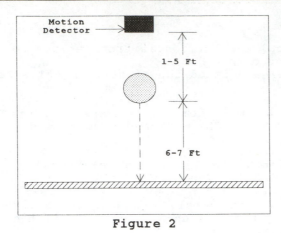

Figure 2

Hold a basketball (or volleyball) above the floor at a height of 6 or 7 feet from the floor to the bottom of the ball. With the calculator attached to the CBL and the motion detector connected to the sonic port of the CBL, turn on the CBL and start the DROPX program. See Figure 3 for CBL hookup. Press the trigger button on the CBL, and within half a second release the ball. Do not throw or push the ball, simply let it fall from rest. The DROPX program records the position of the ball every two hundredths of a second. Do not attempt to drop the ball at exactly the same time that the trigger button is pressed; because readings are being taken every two hundredths of a second, human error will produce inconsistent results.

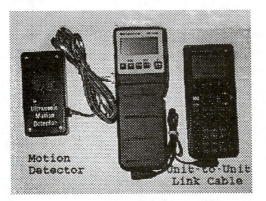

Figure 3

2. The formula for the motion of the ball would be height $= -16x^2 + 8$, if the ball were dropped at time $x = 0$ seconds from a height of $y = 8$ feet. Enter the function $y_1 = -16x^2 + 8$ and overlay its graph on the graph of your data. Notice that the curve does not lie on top of the first arc of the data (see Appendix 1.6). This is because we did not drop the ball at $x = 0$ seconds, and the initial height was not 8 feet. Use vertical and horizontal translations of the function in y_1 until the graph of y_1 lies on top of the first arc of the data. Record the formula for your function in the form $y_1 = -16(x - a)^2 + b$.

    a.   What happened in our physical model at $x = 0$ seconds?

    b.   What happened in our physical model at $x = a$ seconds?

    c.   What is the significance of b in our physical model?

        (Note: you may want to use the trace mode to answer these questions.)

3.   Find a function of the form $y_2 = -16(x - a)^2 + b$ which models the second arc of the data. Record your function.

    a.   What is the physical significance of a?

    b.   What is the physical significance of b?

4.   We now have two separate functions stored in y_1 and y_2. The first, y_1, describes the motion of the ball from the time it is dropped until the time it hits the floor. The second, y_2, describes the motion of the ball after it bounces, until it hits the floor again.

    a.   What is the value of $y_1(0.7)$?

    b.   What does $y_1(0.7)$ represent?

    c.   What is the value of $y_1(2.0)$?

    d.   Does $y_1(2.0)$ make sense in our physical problem? Why or why not?

    e.   What is the value of $y_2(0.7)$?

    f.   Does $y_2(0.7)$ make sense in our physical problem? Why or why not?

    g.   What is the value of $y_2(1.3)$?

    h.   Does $y_2(1.3)$ make sense in our physical problem? Why or why not?

5.   Define a piecewise function which describes the motion of the ball from the time that the ball is dropped until it hits the floor the second time (i.e. include the first two arcs). Use the trace mode to find approximate values for x_0, x_1, and x_2, and record your function in the form:

$$f(x) = \begin{cases} g(x) & , \text{ if } x_0 \le x < x_1 \\ h(x) & , \text{ if } x_1 \le x \le x_2 \end{cases}$$

What do x_0, x_1, and x_2 represent in our physical problem?

6.   To graph a piecewise function on your calculator, you will need to use a different format than the above notation. (See Appendix 1.2). **G1** Print a picture of your piecewise function overlaid on a scatter plot of the original data.

Checklist of calculator graph printouts to be handed in:
☐   **G1**   Print a picture of your piecewise function overlaid on a scatter plot of the original data. (see Appendix 4.1)

Objectives:
    Given the graph of a rational function, find the name of a
    function which has the same x-intercepts, vertical
    asymptotes, and horizontal asymptotes.

Technology:
    TI-82, TI-85, or TI-92 graphing calculator.

Prerequisites:
    Given a rational function, find the x-intercepts, the
    vertical asymptotes, and the horizontal asymptotes.

Overview: The family of rational functions is of the form
$f(x) = \dfrac{P(x)}{Q(x)}$, where P(x) and Q(x) are polynomials. In your first
exposure to functions, you were given an equation and asked to
sketch a graph by finding x-intercepts, vertical asymptotes, and
horizontal asymptotes. The purpose of this project is to begin
with the graph of a rational function, analyze the
characteristics, and find an appropriate name for the function.

Procedures: Find a rational function name for each graph. The
graph of your function should have the following characteristics
in common with the given graph.
1.   Similar behavior as x→ +∞ or x→ −∞.
2.   Same vertical asymptotes.
3.   Same horizontal asymptotes.
4.   Same x-intercepts and the same local behavior around the x-
intercepts.
    After you have guessed a name for the rational function, use
your graphing calculator to graph and compare your "educated
guess" with the given graph (each tick mark on the given graph is
one unit.) See Appendicies 1.1 and 2.5 for function graphing and
window settings. Make any necessary adjustments and continue
until your function is acceptable as defined above. The function
name is not necessarily unique.

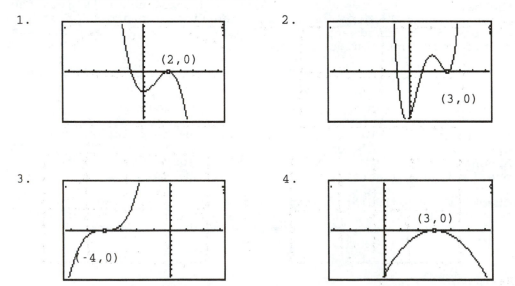

5.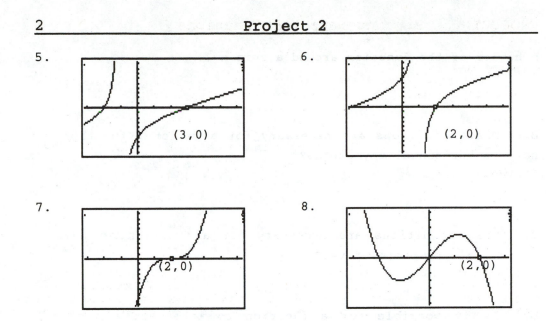

6.

(3,0)

(2,0)

7.

8.

(2,0)

(2,0)

There is a horizontal asymptote of y=1 for each graph in problems 9-13 and y=0 for problem 14.

9.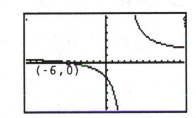

10.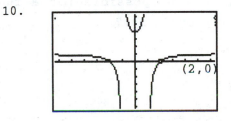

(-6,0)

(2,0)

11.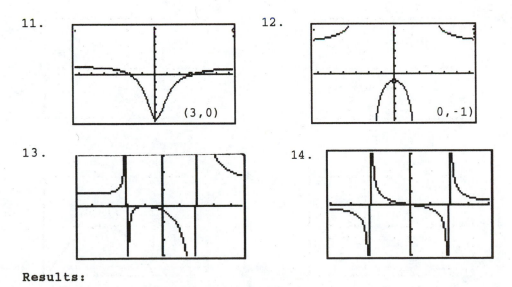

12.

(3,0)

0,-1)

13.

14.

Results:

A.   Discuss how the power of a factor (x-a) determines the local behavior of the function around a zero x=1.

B.   What conditions are necessary for a rational function to have no horizontal asymptotes?

C.   What conditions are necessary for a function to have a horizontal asymptote of y=1?

D.   Is is possible for a function to intersect a vertical asymptote?   Why?

E.   Is it possible for a function to intersect a horizontal asymptote?   Why?

Objectives:
1.   Introduce damped sinusoidal motion.
2.   Model a real-world application with damped sinusoidal motion.
3.   Illustrate the connection between mathematical parameters and physical dimensions.

Technology:
      TI-82, TI-85, or TI-92 graphing calculator.

Prerequisites:
1.   Introduction to amplitude, period, phase shift, and vertical translation of sine and cosine functions.
2.   Introduction to exponential functions.

Overview: The functions $f(t)=ab^t$, where $0 < b < 1$, and $g(t)=\sin(ct)$ can be combined to form a damped sinusoidal function (see Figures 1-3).

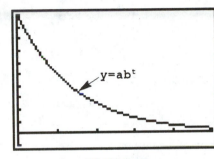

Figure 1

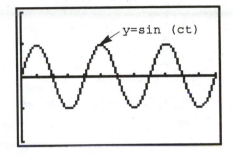

Figure 2

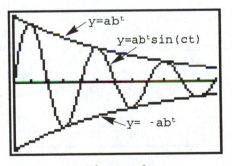

Figure 3

We can think of this new function $h(t) = ab^t\sin(ct)$ as a sine function with a time dependent amplitude of ab^t. We will use a function of this type to describe the motion of a bungee jumper. A bungee jumper has decided to jump off of the Golden Gate Bridge. Not having received the blessings of city officials to jump from the bridge, she must jump from an unmonitored part of the bridge which is only 225 above the surface of the water. She has carefully calculated the length of cord required (she is an excellent physics student) to enable her to skim the surface of the water with her hands on the initial plunge with the cord recoiling before her head dips underwater. Once her hands touch the water, the cord yanks her upward, and when she falls the second time, her hands come within 15 feet of the surface of the water. Her friend has recorded this data from a boat, and has also noted that is took 2.3 seconds from the moment that her hands touched the water to the point in the second fall when her hands were 15 feet from the surface of the water (see Figure 4). The distance from her hands to her feet is about 7 feet as noted in Figure 4.

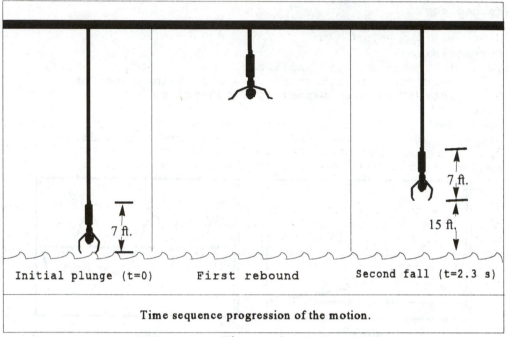

Initial plunge (t=0)   First rebound   Second fall (t=2.3 s)

Time sequence progression of the motion.

Figure 4

After bouncing up and down several times, she finally begins to stabilize with her feet about 50 feet above the surface of the water. We can approximate the height of the jumper's feet above the surface of the water as a function of time using the formula $h(t) = 50 - ab^t \cos(ct)$, where $0 < b < 1$, and $t = 0$ corresponds to the time when her hands touch the water.

Procedures:

1.   Using the fact that h=7 ft when t=0, find a in the formula $h(t) = 50 - ab^t \cos(ct)$.   See Figure 5.

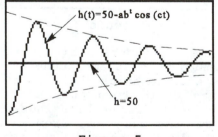

$h(t)=50-ab^t \cos(ct)$

$h=50$

Figure 5

2.   Find c by determining the period of the cosine function from the given data.

3. Using the fact that h=22 ft when t=2.3 sec, find b. Record your function in the form h(t) = 50 - ab^t cos (ct).

4. Graph this function on your calculator to determine if it passes through the points (0, 7) and (2.3, 22). See Appendix 1.1 for function graphing details.

5. Why did we use the "-" sign in the formula h(t) = 50 - ab^t cos (ct)?

6. Why did we use the cosine function instead of the sine function?

7. What is the physical significance of the number 50 with respect to the bungee jumper in the formula for h(t).?

8. The function that you found in Procedure 3 is of the form h(t) = 50 - ab^t cos (ct). Using an appropriate translation, write this function using a sine function in place of the cosine. Use your calculator to graph both functions. If the two graphs are different, adjust your new function and try again. Record the formula for this function.

Objectives:
1.   Develop numerical and graphical techniques for investigating the limit concept.
2.   If a limit exists, investigate short cuts for finding its value.

Technology:
     TI-82, TI-85, or TI-92 graphing calculator.

Prerequisites:
     Working knowledge of function notation.

Overview: The symbols $\lim_{x \to a} f(x) = L$ are read: The limit of the function f as x approaches a is L. The limit concept is used to describe the behavior of f(x) for values of x close to but not equal to a. The meaning of the word "close" varies depending on the context. A function grapher will be used to gain an intuitive understanding of the limit concept and to define what the word close means in this context.

Example 1:   To investigate the limit, $\lim_{x \to 1} \dfrac{x^2 - 8x + 7}{x^2 + x - 2}$, store the function $y = \dfrac{x^2 - 8x + 7}{x^2 + x - 2}$ at the y= menu. Set up a friendly window (see Appendix 2.5) that includes x=1 with $\Delta x = 0.1$. Trace to the point whose x coordinate is 1 and notice that the corresponding y value is not displayed. This trace mode indicates that f(1) is undefined. If the trace mode did not allow you to land exactly on x=1, recheck your friendly window. Set the zoom factors to 10 (see Appendix 3.1) to preserve your friendly window. Zoom in three times (see Appendix 3.2) on the missing point whose x coordinate is 1 and trace to find a pattern for predicting the limit. Tracing produces the following results:

x	.9997	.9998	.9999	1	1.0001	1.0002	1.0003
f(x)	-2.0003	-2.0002	-2.0001	und	-1.9999	-1.9998	-1.9997

As you trace toward 1 from the left of 1, the values of x tend toward 1 and the corresponding y values tend toward -2. Notice a similar behavior pattern when you trace toward x=1 from the right side of 1.

This analysis indicates that as x approaches 1 a prediction for the limit is -2. You can obtain similar results if your function grapher has a table or a table program. Failure to analyze "both sides" of x=1 may lead to incorrect predictions as illustrated in the next example.

Example 2:   To investigate $\lim_{x \to 0} \dfrac{2|x|}{x}$ store $y = \dfrac{2|x|}{x}$ at the y= menu. Set up a friendly window that includes x=0 with $\Delta x = 0.1$. Your graph should look similar to the one in Figure 1. The trace mode should indicate that f(0) is not defined. As you trace the graph for values of x close to zero, what are the y values approaching? If you predict that $\lim_{x \to 0} f(x) = 2$, then you are almost half right. The behavior of y for x close to 0 and x>0 is what you have really described, and we write $\lim_{x \to 0^+} \dfrac{2|x|}{x} = 2$.

The behavior of y for x close to 0 and x<0 is correctly described as $\lim\limits_{x\to 0^-} \dfrac{2|x|}{x} = -2$. Since $\lim\limits_{x\to 0^-} \dfrac{2|x|}{|x|} \neq \lim\limits_{x\to 0^+} \dfrac{2|x|}{|x|}$, $\lim\limits_{x\to 0} \dfrac{2|x|}{x}$ does not exist. Suppose that someone predicted that $\lim\limits_{x\to 0} \dfrac{2|x|}{x} = 0$. This person does not understand the meaning of the word close in this context. For any x, where x is not equal to 0, f(x) is exactly 2 units from zero. By "close" we mean arbitrarily close. It is impossible to pick values of x close to 0 that produce corresponding y values that are arbitrarily close to 0. If the behavior of y=f(x) for x close to a and x>a is different from the behavior when x is close to a and x<a, then the limit does not exist.

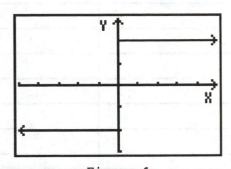

Figure 1

I.   Investigate the following limits by zooming in and tracing or by using a table of numbers to determine if the limit exists. If it does, predict its value and record this information in Table 1. Fill in the function column with exact values. Set up a friendly window with Δx=0.1

1. $\lim\limits_{x\to 2} (5x+1)$

2. $\lim\limits_{x\to 2} \dfrac{x+3}{x-2}$

3. $\lim\limits_{x\to -2} \dfrac{(x^2+4)}{x+2}$

4. $\lim\limits_{x\to 3} \dfrac{|x-3|}{x-3}$

5. $\lim\limits_{x\to 1} \dfrac{2x+1}{x+1}$

6. $\lim\limits_{x\to 1} 2x^2+3x-1$

7. $\lim\limits_{x\to -.1} \dfrac{|x|}{x}$

8. $\lim\limits_{x\to \frac{1}{3}} \dfrac{3x^2+2x-1}{9x^2-1}$

9. $\lim\limits_{x\to \frac{1}{2}} \dfrac{2x^2+5x-3}{2x^2-3x+1}$

f(x)	a	f(a)	$\lim\limits_{x \to a} f(x)$
1)			
2)			
3)			
4)			
5)			
6)			
7)			
8)			
9)			

Table 1

Compare the results in the columns for f(a) and $\lim\limits_{x \to a} f(x)$. 10.

Conjecture when and how to find the value of $\lim\limits_{x \to a} f(x)$ if the limit exists.

II.   Investigate $\lim\limits_{x \to a} F(x)$ where $F(x) = \dfrac{P(x)}{Q(x)}$. Do not simplify any functions prior to filling in Table 2. Graph and zoom in enough times to predict whether or not the limit exists, and if it does, give its value. Do not approximate Q(a), P(a), or F(a), but compute their exact values. Note that $F(x) = \dfrac{P(x)}{Q(x)}$. Fill in Table 2 completely.

1. $\lim\limits_{x \to -2} \dfrac{x^2 - 1}{x + 2}$

2. $\lim\limits_{x \to 2} 3x^2 + 2x$

3. $\lim\limits_{x \to 0} \dfrac{x^4}{x^5 + 5x^4}$

4. $\lim\limits_{x \to 2} \dfrac{(x-2)(x+5)}{x-2}$

5. $\lim\limits_{x \to -3} \dfrac{x^2 + 5x + 6}{x^2 + 6x + 9}$

6. $\lim\limits_{x \to 2} \dfrac{x+5}{x-2}$

7. $\lim\limits_{x \to 3} \dfrac{x^3 - 6x^2 + 9x}{x^2 - 2x - 3}$

8. $\lim\limits_{x \to 1} \dfrac{(x-3)(x-1)}{(x-1)(x+2)}$

9. $\lim\limits_{x \to 1} \dfrac{x-5}{x+7}$

	$F(x)$	a	$P(a)$	$Q(a)$	$F(a)$	$\lim_{x \to a} F(x)$
1)						
2)						
3)						
4)						
5)						
6)						
7)						
8)						
9)						

Table 2

Analyze Table 2 and answer the following questions.

1.   If $F(a)$ is undefined, does it follow that $\lim_{x \to a} F(x)$ does not exist?

2.   When $F(a)$ is undefined, under what conditions does $\lim_{x \to a} F(x)$ exist? You may want to simplify $F(x)$ to help you arrive at a conclusion.

3.   Summarize your observations with a conjecture for when $\lim_{x \to a} F(x)$ does or does not exist, where $F(x)$ is a rational function.

4.   Under certain conditions, can you find a shortcut for finding $\lim_{x \to a} F(x)$? If your answer is yes, describe your shortcut.

Please note: some of the conclusions that you made int his project only apply to special types of functions. These concepts will be further investigated later in our discussion of continuity.

Objectives:

1.   Provide evidence to support a conjecture that $\lim_{x \to a} f(x) = L$.

2.   Provide evidence to support a conjecture that $\lim_{x \to a} f(x) \neq L$.

Technology:
     TI-82, TI-85, or TI-92 graphing calculator.

Prerequisites:
     None

Overview: The symbols $\lim_{x \to a} f(x) = L$ intuitively means that values
of f(x) are close to L whenever x is close to a but not equal to
a.   The word close is too ambiguous to define the mathematical
concept of a limit.   The meaning of the word close will be
illustrated by tolerances for x and y. $\lim_{x \to a} f(x) = L$ means that for
any y tolerance ϵ about L there is a x tolerance, δ, about x=a
such that if   a-δ < x < a+δ (x is within a δ tolerance of a),
then L-ϵ < y < L+ϵ (y is within an ϵ tolerance of L).   See Figure
1.

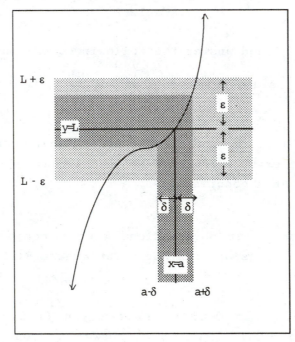

Figure 1

Example 1: To support the claim that $\lim_{x \to 1} \dfrac{x^2 - 8x + 7}{x^2 + x - 2} = -2$, set up a

friendly window including x=1 with Δx=0.1.   (see Appendix 2.5).
Set Ymin=-3 and Ymax=2.   Your graph of $y = \dfrac{x^2 - 8x + 7}{x^2 + x - 2}$ should be
similar to the one in Figure 1.   Suppose someone gives you an ϵ
tolerance about y=-2 of 0.1 and then challenges you to find a δ
tolerance about x=1 such that for x in the interval 1-δ < x < 1+δ,
the corresponding y values will be in the interval -2-.01<y<-2+.01

or -2.01<y<-1.99. The zoom-in and the trace features can be used to find an appropriate δ. Graph $y=\dfrac{x^2-8x+7}{x^2+x-2}$, y=-2.01 and y=-1.99 on the same screen with a friendly window including x=1 with Δx=.1. Your graph should be similar to the one in Figure 2.

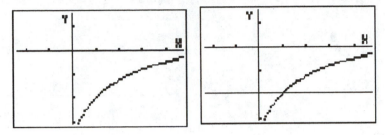

Figure 2   Figure 3

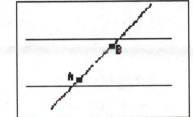

Figure 4

In Figure 3 it is difficult to distinguish between the graphs of the horizontal lines y=-2.01 and y=-1.99. We need a snapshot of x values close to 1 and y values close to -2.

Set the zoom factors to 10 and zoom in twice on the point whose x-coordinate is 1 and the y-coordinate is as close as possible to -2. Your graph should be similar to the one in Figure 4. The graphs of y=-2.01 and y=-1.99 should be distinct and you should be able to see a "hole" in the graph of $y=\dfrac{x^2-8x+7}{x^2+x-2}$. Trace to find two points A and B that lie between the two lines and are on opposite sides of the "hole". Acceptable coordinates for A and B are (0.991, -2.009027) and (1.009, -1.991027). Restrict your attention to points on the graph of $y=\dfrac{x^2-8x+7}{x^2+x-2}$ that lie between points A and B. Restricting the x values to the interval (.991, 1.009) assures that the ε-challenge is met for ε=.01. In other words, the y-coordinate of any such point is between y=-2.01 and y=-1.99 (see Figure 5). This choice of points produces a suitable δ for x about x=1.

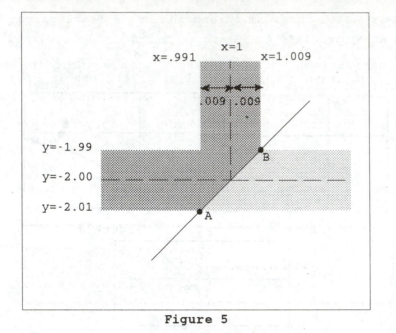

Figure 5

Notice that .991 < x < 1.009 is the same as   1-.009 < x < 1+.009
so an appropriate δ tolerance for x is 0.009.   Your choice for δ
may differ.   It is possible that your choices for A and B yield
values of x that are not equidistant from 1 (see Figure 6).   If
the distances from 1 differ as illustrated in Figure 6, choose δ
to be the smaller distance.   In our particular example, the
distances are .009 and .005, so we will choose δ = .005.   The
reason for choosing the smaller value for δ is illustrated in
Figure 6.   The smaller value determines a narrower strip which
must meet the є challenge also.

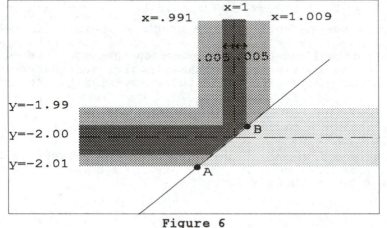

Figure 6

In summary, to support the statement, $\lim_{x \to a} f(x) = L$, for any given
є > 0, graph y=f(x), y=L-є and y=L+є.   Then zoom in on the
coordinates (x,L) enough times to distinguish the tolerance lines
Y= L+є and Y= L-є.   Use the trace mode to find acceptable x
points A and B.   From those points determine an acceptable δ
tolerance.

Example 2: Suppose a student claims that for the function $y = \frac{2|x|}{x}$, shown in Figure 7, $\lim_{x \to 0} \frac{2|x|}{x} = 2$. This claim is false. In order to prove that the limit is **not** 2, we must choose a value of ϵ such that no δ-tolerance can be found about x=0 which will guarantee that all values of x with this δ-tolerance will have corresponding y values within the ϵ-tolerance about y=2. In order to find an appropriate value of ϵ, we will study the behavior of the function on both sides of x=0. Since y=2 for all x>0, and y=-2 for all x<0, any value of ϵ which is less than 4 will work. For example, suppose we choose ϵ=1. The corresponding ϵ-tolerance about y=2 is shown in Figure 8. This choice of ϵ will "over-challenge" anyone looking for an acceptable δ-tolerance, since there is no point A, to the left of x=0 which is within the ϵ-tolerance.

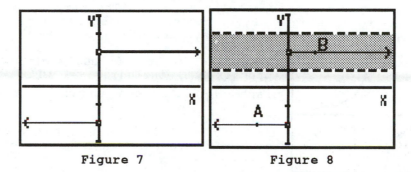

Figure 7   Figure 8

Since y=2 whenever x>0, we say that the right-hand limit is 2. Since y=-2 whenever x<0, we say that the left-hand limit is -2. Because of this different "one-sided" behavior, $\lim_{x \to 0} \frac{2|x|}{x}$ does not exist.

Procedures: For each of the following, determine if the limit statement is true or false. If the statement is true, your challenge is to find a δ-value for ϵ = 0.01. In this case, find the coordinates of points A and B and use Example 1 as a guide to find δ. If the statement is false, give a value of ϵ that will over-challenge anyone looking for an acceptable δ. In this case, print out a graph (see Appendix 4.1) and record your ϵ value.

1. $\lim_{x \to 1} (3x^2 + 1) = 4$ 2. $\lim_{x \to 0} \frac{|x|}{2x} = \frac{1}{2}$

3. $\lim_{x \to 1} \frac{x^2 - 4}{x - 4} = 1$ 4. $\lim_{x \to 1} \frac{2x^2 + x - 3}{x - 1} = 5$

5. $\lim_{x \to 1} (2x - 3) = 0$ 6. $\lim_{x \to 2} \frac{x}{x - 2} = 1$

Checklist of calculator graph printouts to be handed in:

☐   For any false statements, print out a graph showing that the value of ϵ will over-challenge anyone looking for an acceptable δ.

Objectives:
    Expand the limit concept to describe the local behavior of
    a function around a vertical asymptote.

Technology:
    TI-82, TI-85, or TI-92 graphing calculator.

Prerequisites:
    Introduction to limits.

Overview: The limit concept is used to describe the behavior of
the dependent variable, y, for a specified behavior of the
independent variable, x.   The statement $\lim_{x \to a} f(x) = L$ implies that

as the independent variable x gets "close" to a, the
corresponding y-values are "close" to L.   The purpose of this
project is to expand the limit concept to describe a different
kind of behavior for the dependent variable.

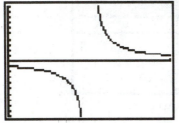

Example 1: Set up a friendly window (see
Appendix 2.5) including x=1 with Δx=.1,
graph $f(x) = \dfrac{1}{x-1}$, and compare your graph
with the one in Figure 1.   This graph
does not provide enough information to
completely describe the behavior of y as
x gets close to 1.

Figure 1

Notice that the function is undefined at x = 1.   The information
in Table 1 more accurately describes the behavior of y as x gets
close to 1, for x > 1.   The dependent variable y does not appear
to be getting close to any number.   In fact, it appears that the
values of y are increasing without bound.   The symbolism
$\lim_{x \to 1^+} f(x) = +\infty$ indicates that as $x \to 1^+$ (x approaches 1 from the

right) the values of y increase without bound.

x	2	1.5	1.1	1.01	1.001	1.0001
y	1	2	10	100	1000	10000

Table 1

The data in Table 2 provides some insight into the behavior of y
as x approaches 1 from the left.   The dependent variable, y, is

x	0	.9	.99	.999	.9999
y	-1	-10	-100	-1000	-10000

Table 2

decreasing   without   bound.   These   words   are   described
symbolically by: $\lim_{x \to 1^-} f(x) = -\infty$.   The symbols "+∞" and "-∞" are not

treated values of the limit, but are used as symbols to describe
the   behavior   of   the   dependent   variable.   Because   of   this

behavior, the vertical line x = 1 is called a vertical asymptote. See Figure 2.

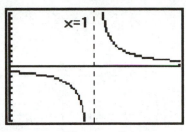

Figure 2

Example 2: Set up a friendly window including x=1 with Δx=.1 and graph $f(x) = \frac{3x^2 - 5x + 2}{x - 1}$.

See Figure 3. This function is undefined at x = 1. If you can't see the "hole", trace over to x = 1 and notice that the y coordinate is not displayed. This indicates that y = f(1) is undefined. This is an example of a function which is undefined at x = 1 but has no vertical asymptote there. Later, after working through several problems, you will be asked to make a conjecture about the existence of a vertical asymptote at x = a, for various types of functions, when f(a) is undefined.

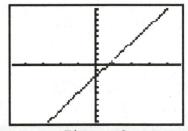

Figure 3

Example 3: Sometimes when you use a graphing calculator to graph a function with a vertical asymptote, it will appear as though a vertical asymptote is also graphed. The graph of $y = \frac{1}{x - 1}$ with a standard window is displayed in Figure 4. The graph appears to contain the vertical asymptote, x=1. The way your calculator constructs a graph is by joining two adjacent points with a line segment. The segment that appears to be vertical is a connecting segment and not a vertical asymptote and is not a part of the graph.

Figure 4

Procedures: For each function, determine if there are one or more vertical asymptotes. If so, list them and use limit notation such as

$$\lim_{x \to a^+} f(x) = -\infty,$$

$$\lim_{x \to a^+} f(x) = +\infty,$$

$$\lim_{x \to a^-} f(x) = -\infty,$$

or $$\lim_{x \to a^-} f(x) = +\infty$$

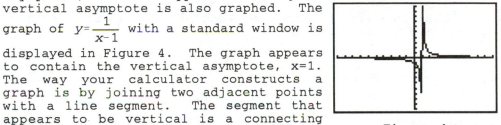

to describe the behavior of the function around each value of x for which a vertical asymptote exists. Do not approximate the x coordinates of vertical asymptotes but give exact values if possible. If an exact value is not possible, estimate it to 4 decimal places.

1. $f(x) = \dfrac{x^2 + 5x + 6}{x - 2}$

2. $f(x) = \dfrac{(x + 3)(x + 2)}{(x + 3)^2}$

3. $f(x) = \dfrac{x + 3}{(2 - x)(x + 1)}$

4. $f(x) = \dfrac{x + 3}{(2 - x)(x + 1)^2}$

5. $f(x) = \dfrac{x^2 + 6x + 9}{x^2 + x + 1}$

6. $f(x) = \dfrac{x + 2}{x^2 + 1}$

7. $f(x) = \dfrac{x + 1}{x^2 + x - 3}$

8. $f(x) = \dfrac{x^2 - 1}{x^3 - 3x + 3}$

9. $f(x) = \dfrac{x^2 - 1}{x + 1}$

10. $f(x) = \dfrac{x^2 + 2x - 3}{x^2 - 6x + 5}$

11. $f(x) = \dfrac{x^2 + 1}{|x + 1|}$

12. $f(x) = \dfrac{\sin x}{x}$

13.   Describe any patterns between equations and vertical asymptotes for the functions in 1-12 above.

Objectives:
1.   Expand the limit concept to describe "end" behavior of a
      function.
2.   Find horizontal asymptotes for rational functions.

Technology:
      TI-82, TI-85, or TI-92 graphing calculator, and program TAB
      (for TI-85 only).

Prerequisites:
      Introduction to limits.

Overview: The limit concept is used to describe the local
behavior of a function around some number 'a'.   The purpose of
this project is to expand the limit concept to describe the end
behavior of a function.   The statement $\lim\limits_{x \to a} f(x) = L$ indicates that

if x is close to a then f(x) is close to L.   This idea helps you
to visualize to some extent the way the dependent variable
changes when x is close to a.   This is called local behavior of
f(x) around a.

The purpose of this project is to develop the intuition and
symbolism to describe the behavior y when the values of x
increase or decreases without bound.   This y-behavior is referred
to as end behavior.

Example 1:   The graph of $f(x) = (1.1)^x + 1$
in Figure 1 gives a view of the
behavior of y=f(x) for x between -10
and 10.   It is not evident how to
describe the end behavior of this
function.   As   the   values   of   x
increase,   the   values   of   y   also
increase   and   as   the   values   of   x
decrease, so do the values of y.   We
want to determine if y has a limiting
value as x increases without bound

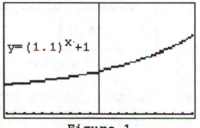

Figure 1

and, if it doesn't, find a way to describe the "behavior" of the
corresponding values of y.   Table 1 describes the function's
behavior for larger values of x.   See Appendix 5.1 for details on
using tables.

x	0.0	10.0	20.0	30.0	40.0	50.0	100
y	2.0	3.6	7.7	18.4	46.3	118.4	13782

Table 1

Based on Table 1, it appears as if the values of y increase without bound as x increases without bound. To view the graph in Figure 2 for x=70 to x=100, the range of y is set from 1,000 to 15,000. If you choose larger x-values, the y-values continue to grow without bound. This is represented by the symbolic statement

$$\lim_{x \to +\infty} f(x) = +\infty.$$

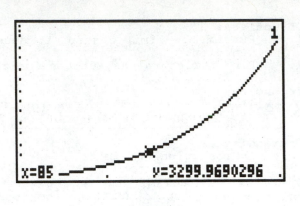

Figure 2

The graphs in Figures 3 through 5 illustrate the language and symbolism for describing right and left end behavior of each function.

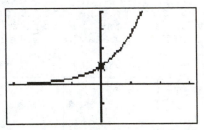

Figure 3

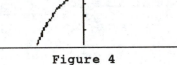

Figure 4

Right end behavior:
as x→+∞, y increases
without bound (y→+∞)

$$\lim_{x \to +\infty} f(x) = +\infty$$

Right end behavior
as x→+∞, y approaches 1
(y→1)

$$\lim_{x \to +\infty} f(x) = 1$$

Left end behavior:
as x→-∞, y approaches
zero (y→0)

$$\lim_{x \to -\infty} f(x) = 0$$

Left end behavior
as x→-∞, y decreases
without bound (y→-∞)

$$\lim_{x \to -\infty} f(x) = -\infty$$

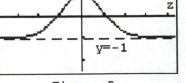

Figure 5

Right end behavior:
as x→+∞, y approaches -1
(y→-1)

$$\lim_{x \to +\infty} f(x) = -1$$

Left end behavior:
as x→-∞, y approaches -1
(y→-1)

$$\lim_{x \to -\infty} f(x) = -1$$

Example 2: Consider the function $y=\dfrac{2x^2+13}{x^2+1}$. The values of x in Table 1 are chosen to determine the behavior of y as x increases without bound.

x	10	20	80	100	200	400
y	2.1089	2.0274	2.0017	2.0011	2.0003	2.0001

Table 1

The information in this table supports the conjecture that $\lim\limits_{x\to+\infty}\dfrac{2x^2+13}{x^2+1}=2$. The graph of y=2 overlaid on the graph of $y=\dfrac{2x^2+13}{x^2+1}$ is displayed in Figure 6.

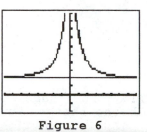

Figure 6

This graphical view also supports the conjecture that $\lim\limits_{x\to+\infty}\dfrac{2x^2+13}{x^2+1}=2$. The graph of y=2 and $y=\dfrac{2x^2+13}{x^2+1}$ is displayed in Figure 7 for x ranging from -20 to 1. This graph leads to a conjecture that $\lim\limits_{x\to-\infty}\dfrac{2x^2+13}{x^2+1}=2$.

Figure 7

Prepare a table with appropriate values of x to support this conjecture. Based on the limit information $\lim\limits_{x\to+\infty} f(x)=2$ and $\lim\limits_{x\to-\infty} f(x)=2$ the line y=2 is called a horizontal asymptote.

Work through the following exploratory activities. Graph and observe!

Procedures:
I. Graph each of the following functions and investigate the left and right end behavior. For each of the following functions, determine whether or not a horizontal asymptote exists. If a horizontal asymptote exists then write its equation.

1. $y=\dfrac{3x^2+1}{x^2+1}$

2. $y=\dfrac{3x+1}{x^3+1}$

3. $y=\dfrac{7x+1}{2x^2+1}$

4. $y=\dfrac{6x^2+1}{3x^2+7}$

5. $y=\dfrac{x^3+1}{x^2+1}$

6. $y=\dfrac{x^4+1}{7x^3+1}$

7. $y = \dfrac{2x^5 + 1}{x^4 + 2}$ 8. $y = \dfrac{6x^3 + x + 1}{2x^3 + 5}$

9.   By looking for patterns between functions and asymptotes, write a paragraph on how you can tell from the defining equation of a rational function whether it has a horizontal asymptote. If it has one, tell how to find it without graphing it or using a table of values.

II.   For these exercises, $f(x) = 6x^2 + 5$, $g(x) = 2x^3$, $h(x) = 2x^2 + 4x + 6$. Choose the best response.

1.   If $y \to +\infty$ when $x \to +\infty$ then $y=$

   a) $\dfrac{f(x)}{g(X)}$ b) $\dfrac{h(x)}{f(x)}$ c) $\dfrac{g(x)}{f(x)}$ d) $\dfrac{h(x)}{g(x)}$

2.   If $y \to 0$ when $x \to +\infty$ then $y=$

   a) $\dfrac{f(x)}{g(X)}$ b) $\dfrac{h(x)}{g(x)}$ c) $\dfrac{g(x)}{h(x)}$ d) $\dfrac{f(x)}{h(x)}$

   e)   both a and b are correct

3.   If $y \to \dfrac{1}{3}$ when $x \to +\infty$ then $y=$

   a) $\dfrac{f(x)}{h(x)}$ b) $\dfrac{g(x)}{f(x)}$ c) $\dfrac{h(x)}{f(x)}$ d) $\dfrac{f(x)}{g(X)}$

4.   If $y \to 4$ when $x \to +\infty$ then $y=$

   a) $f(x) + f(x)$ b) $\dfrac{f(x) + h(x)}{h(x)}$ c) $g(x) \times h(x)$

   d) $g(x) + h(x)$

III. Based on the given partial graph, choose the best model.

1.

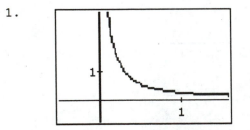

a) $y = \dfrac{x^3 + x}{x^3 + 1}$ b) $y = \dfrac{x^2 + 2x}{6x^3 + 4x^2}$

c) $y = \dfrac{x^5 - 1}{x^4 + 1}$ d) $y = x^2 + 3x + 1$

2.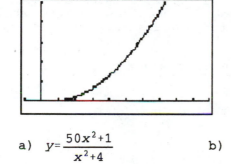

a) $y = \dfrac{x^2 + 1}{x^4 + 1}$ b) $y = \dfrac{8x^6 + x^2}{4x^6 + 5}$

c) $y = \dfrac{2x^3 + 1}{x^2 + 1}$ d) $y = \dfrac{x^2 + 4}{x^2 + 2}$

3.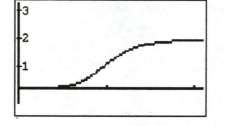

a) $y = \dfrac{50x^2 + 1}{x^2 + 4}$ b) $y = \dfrac{100x + 5}{x + 2}$

c) $y = \dfrac{x^4 + x^2}{x^2 + 5}$ d) $y = \dfrac{x^2 + 5}{x^4 + x^2 + 1}$

Objectives:
1.   Determine if a function is continuous at a point $(a, f(a))$
2.   Give at least three graphical descriptions of how a function can be discontinuous at a point $(a, f(a))$
3.   Describe algebraically different instances for which f is discontinuous at a point $(a, f(a))$

Technology:
    TI-82, TI-85, or TI-92 graphing calculator.

Prerequisites:
1.   An introduction to limits and one sided limits.
2.   An introduction to piecewise functions (Procedure II only).

Overview: The concept of a limit of a function can be investigated by creating and analyzing a table, analyzing a graph, or by using algebraic techniques. A function f is continuous at a number a if and only if:

1. $\lim\limits_{x \to a} f(x)$ and f(a) are both defined and are real numbers.

2. $\lim\limits_{x \to a} f(x) = f(a)$.

The primary purpose of this project is to help you gain a feel both graphically and numerically for what it means to say that a function f is continuous at x=a.   Sometimes insight about a property can be obtained by studying instances for which the property does not hold.

A function f can be discontinuous at x=a for any one of three reasons:
1.   f(a) is undefined

2. $\lim\limits_{x \to a} f(x)$ does not exist

3. $\lim\limits_{x \to a} f(x) \neq f(a)$

Procedures:
I.   Set up a friendly window (see Appendix 2.5) with $\Delta x = .1$ centered about x=0. The friendly window is necessary to graphically determine where a function is undefined.   Find $\lim\limits_{x \to a} f(x)$ and f(a).   Determine if the function is continuous at

a.   If it is not continuous at a, give a specific reason from the above list.

   1. $f(x) = -3x^2 + 2$, a=2

   2. $f(x) = \dfrac{2x^2 - x - 6}{x - 2}$, a=2

   3. $f(x) = \dfrac{3x^3 - 7x^2 - 12x + 28}{5x^2 + 3x + 2}$, a=2

   4. $f(x) = \dfrac{x^2 - x + 1}{x - 2}$, a=2

5. $f(x) = \dfrac{|x^2 - x - 2|}{x - 2}$, a=2

6. $f(x) = \dfrac{x - 3}{x + 2}$, a=-2

II. Use the same friendly window as in Procedure I. See Appendix 1.2 for techniques of graphing piecewise functions. Find $\lim\limits_{x \to a} f(x)$ and f(a). Determine if the function is continuous

at a. If it is not continuous at a, give a specific reason from the above list.

1. $f(x) = \begin{cases} 3x - 2 & ,\ x \le 2 \\ x^2 & ,\ x > 2 \end{cases}$, a=2

2. $f(x) = \begin{cases} 3x - 1 & ,\ x \le 2 \\ 5x - 1 & ,\ x > 2 \end{cases}$, a=2

3. $f(x) = \begin{cases} 3x^2 - 4x + 1 & ,\ x \ge 1.5 \\ 3x - 2 & ,\ x < 1.5 \end{cases}$, a=1.5

4. $f(x) = \begin{cases} 2x^2 - 3x + 2 & ,\ x \ge .8 \\ 5x - 3 & ,\ x < .8 \end{cases}$, a=.8

5. $f(x) = \begin{cases} 2x + 1 & ,\ x \ne 3 \\ 2 & ,\ x = 3 \end{cases}$

III. Analyze each of the functions which are discontinuous in Procedures I and II. For each of the four cases below, choose a function that corresponds to the given description and print out its graph. In each case write a graphical interpretation of the discontinuity.

1. $\lim\limits_{x \to a} f(x)$ exist and f(a) is undefined.

2. $\lim\limits_{x \to a} f(x)$ exist and f(a) is defined but

       $\lim\limits_{x \to a} f(x) \ne f(a)$.

3. $\lim\limits_{x \to a} f(x)$ does not exist and f(a) is defined.

4. $\lim\limits_{x \to a} f(x)$ does not exist and $f(a)$ is undefined.

IV.   For this part, use the trig option of the zoom mode to set up your window (see Appendix 2.2). Your calculator should be in the radian mode. For each of the following functions record the set of numbers for which the function is discontinuous over the interval $(-2\pi, 2\pi)$. Do not rely exclusively on the graph that your calculator produces, but consider also the three ways that a function can be discontinuous.

     1. $y = \cos x$

     2. $y = \csc x$

     3. $y = \tan x$

     4. $y = \dfrac{1 - \cos^2 x}{\sin x}$

     5. $y = \dfrac{\sin(2x)}{\sin x}$

     6. $y = \dfrac{\cos x}{\sin(2x)}$

     7. $y = \sin \dfrac{1}{x}$

     8. $y = x \sin x$

Objectives:
    Investigate the relationship between increasing and decreasing functions and the derivative of a function.

Technology:
    TI-82, TI-85 (CBL compatable version), or TI-92 graphing calculator, CBL, CBL motion detector, and program HIKEREL.

Prerequisites:
    Definition of the derivative.

Overview: A function is increasing over an interval I if tracing from left to right on I produces increasing values of y. It is decreasing over I if tracing on I from left to right produces decreasing values of y.   The function shown in Figure 1 is decreasing over intervals I_1 and I_3 and is increasing over I_2.

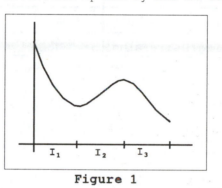

Figure 1

Procedures:
I.   Connect the motion detector to the sonic port of the CBL, and connect the CBL to the calculator with a unit-to-unit link cable.   See Figure 2 for CBL hookup.

Figure 2

Set the motion detector on a table so that the beam faces the "hiker".   The motion detector can only detect rectilinear motion along the line of the beam, so the "hiker" must stay in its path at all times.
1.   Run program HIKEREL and when you are ready to start collecting data, press **[ENTER]** on the calculator and the **[TRIGGER]** on the CBL.   Have a member of your group walk in a

manner such that the graph of f(t) (distance) vs. time is an increasing function. **G1** Printout a scatter plot of f(t) vs. time and y=f'(t) vs. time on the same screen. Time, f(t) and f'(t) are stored in lists L1, L2 and L3 respectively. See Appendix 1.5 for details on scatter plots and Appendix 4.1 for information on obtaining a printout. Analyze the graphs of f vs. time and f' vs. time and fill in the first line of Table 1.

2. Repeat the experiment in #1 and print out the appropriate graphs, **G2**, but this time have a member of your group walk in such a manner that the graph defines a decreasing function.

3. Repeat the experiment in #1 and print out the appropriate graphs, **G3**. This time, have a member of your group walk in such a manner that the graph for the first half of the time is increasing and during the second half the graph is decreasing.

Behavior of f(t)	Sign of f'(t)
1. Increasing	
2. Decreasing	
3. Increasing/Decreasing	

Table 1

4. Analyze Table 1 and write the result in the "if p then q" form, where p describes the sign of the derivative and q describes the behavior of the function.

II. Assume that f(t) vs. time and f'(t) vs. time data is collected for a person moving along a line in the range of a motion detector.

1. Suppose the data analysis indicates that f'(t) is positive over a time interval. Describe the action of the hiker with respect to the motion detector over this time interval.

2. Suppose the data analysis indicates that f'(t) is negative over a time interval. Describe the motion of the hiker with respect to the motion detector over this time interval.

3. Suppose the data analysis indicates that f'(t)=0 over a time interval. Describe the motion of the hiker with respect to the motion detector over this time interval.

Checklist of Calculator Graph Printouts to be Handed in:
☐ **G1** Scatter plot of distance vs. time and f'(t) vs. time for an increasing function.
☐ **G2** Scatter plot of distance vs. time and f'(t) vs. time for a decreasing function.
☐ **G3** Scatter plot of distance vs. time and f'(t) vs. time for a function which is increasing and decreasing.

Objectives:
   Investigate the relationship between concavity of a function
   and the second derivative of the function.

Technology:
   TI-82, TI-85 (CBL compatable version), or TI-92 graphing
   calculator, CBL, CBL motion detector, and program HIKEREL.

Prerequisites:
   Definition of the second derivative.

Overview: A function is concave up on an interval (a,b) if the
curve is above the tangent line at (c,f(c)) where a<c<b (see
Figure 1). A function is concave down over an interval (a,b) if
the curve is below the tangent line at (c,f(c)) where a<c<b (see
Figure 2).

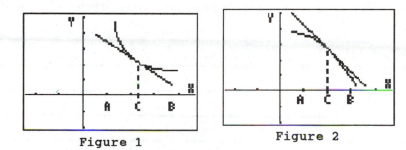

Figure 1   Figure 2

Procedures:
Connect the motion detector to the sonic port of the CBL, and
connect the CBL to the calculator with a unit-to-unit link cable.
See Figure 3 for CBL hookup.

Figure 3

Set the motion detector on a table so that the beam faces the
"hiker". The beam can only detect rectilinear motion along the
line of the beam, so the "hiker" must stay in its path at all
times. For each of the following four cases, run program HIKEREL
and when a member of your group is ready to produce motion data,
press **[ENTER]** on the calculator and then **[TRIGGER]** on the CBL.

1.   Have a member of your group walk in such a manner that the
graph of f(t) (distance) vs. time is concave up over the time
interval for which data is collected. **G1** Print out scatter plots
of f(t) vs. time and f″(t) vs. time on the same screen. See
Appendix 1.5 for details on scatter plots and Appendix 4.1 for

obtaining a printout.   **G2** Print out scatter plots of f(t) vs. time and f'(t) vs. time on the same screen. Analyze the graphs of f"(x) and f(x) and fill in Table 1 with a brief description of the behavior of f'(t) and the sign of f"(t).

2.   Repeat the direction in #1 and print out the appropriate scatter plots, **G3** and **G4**, but this time have a member of your team walk in such a manner that the graph of f(t) vs. time is concave down over the time interval.

3.   Repeat the directions in #1 and print out the appropriate scatter plots, **G5** and **G6**, but this time have a member walk in such a manner that initial part of the graph of f(t) vs. time is concave up and the second part of this graph is concave down.

4.   Repeat the directions again and print out the appropriate scatter plots, **G7** and **G8**, but this time have a member of your group walk so that the initial part of the graph of f(t) vs. time is concave down and the second part of this graph is concave down.

Behavior of f(t)	Behavior of f'(t)	Sign of f"(t)
1. CU		
2. CD		
3. CU/CD		
4. CD/CU		

Table 1

5.   Analyze Table 1 and describe the relationship between f(t) and f'(t).

6.   Describe the relationship between f(t) and f"(t).

7.   Describe the relationship between f'(t) and f"(t).

8.   What can you say about the behavior of f'(t) on any interval where the sign of f"(t) doesn't change?

Checklist of calculator graph printouts to be handed in:

☐   **G1**   Print out scatter plots of f(t) vs. time and f"(t) vs. time on the same screen for concave up.

☐   **G2**   Print out scatter plots of f(t) vs. time and f'(t) vs. time on the same screen for concave up.

☐   **G3**   Print out scatter plots of f(t) vs. time and f"(t) vs. time on the same screen for concave down.

☐   **G4**   Print out scatter plots of f(t) vs. time and f'(t) vs. time on the same screen for concave down.

☐   **G5**   Print out scatter plots of f(t) vs. time and f"(t) vs. time on the same screen for concave up/concave down.

☐   **G6**   Print out scatter plots of f(t) vs. time and f'(t) vs. time on the same screen for concave up/concave down.

☐   **G7**   Print out scatter plots of f(t) vs. time and f"(t) vs. time on the same screen for concave down/concave up.

☐   **G8**   Print out scatter plots of f(t) vs. time and f'(t) vs. time on the same screen for concave down/concave up.

Objectives:
1. Use the symmetric difference quotient to graphically predict the derivative formulas for the sine, cosine, and exponential functions.
2. Use the symmetric difference quotient to predict the derivative of exponential and trigonometric functions when composition is involved.

Technology:
   TI-82, TI-85, or TI-92 graphing calculator.

Prerequisites:
   Introduction to the derivative.

Overview: Given a function $f(x)$, $f'(x)$ can be defined by either $\lim\limits_{h \to 0} \dfrac{f(x+h) - f(x)}{h}$ or $\lim\limits_{h \to 0} \dfrac{f(x+h) - f(x-h)}{2h}$.   For a given value of h, the symmetric difference quotient $\dfrac{f(x+h) - f(x-h)}{2h}$ is usually a better approximation of $f'(x)$ than the difference quotient $\dfrac{f(x+h) - f(x)}{h}$.   Data will be generated to represent $f'(x)$ by using the symmetric different quotient $\dfrac{f(x+.001) - f(x-.001)}{.002}$.   A prediction for $f'(x)$ will then be made by fitting a curve to this data.   The purpose of this project is to use curve fitting and symmetric difference quotients to discover some results about derivatives.

Example 1:   The symmetric difference quotient for $f(x) = \dfrac{1}{x}$ is

$$g(x) = \frac{\dfrac{1}{x+.001} - \dfrac{1}{x-.001}}{.002}$$

or $$g(x) = \frac{(x-.001) - (x+.001)}{.002\,(x+.001)\,(x-.001)}.$$

$$g(x) = \frac{-1}{(x+.001)\,(x-.001)}$$

Graph $g(x)$ and compare your results with the graph in Figure 1. This graph of the symmetric difference quotient for h=.001 is an excellent approximation to the graph of the derivatives of f. We will attempt to find a name for the derivative, f', by finding a name for this graph of $g(x)$.

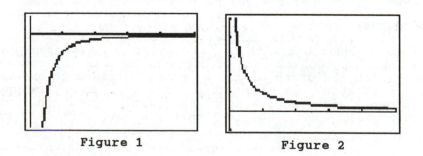

| Figure 1 | Figure 2 |

Comparing the graph of $g(x)$ in Figure 1 with the graph of $f(x) = \dfrac{1}{x}$

in Figure 2, it appears that the graph of y=g(x) is a reflection about the x-axis of the graph of $y=\frac{1}{x}$. The graphs of $y=-\frac{1}{x}$ and y=g(x) in Figure 3 are similar but clearly not the same. Next, let's try $y=-\frac{1}{x^2}$ as a fit for f'(x). The graphs of y=g(x) and $y=-\frac{1}{x^2}$ are overlaid in Figure 4.

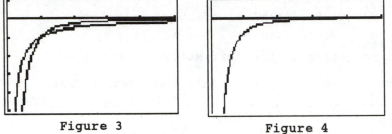

Figure 3   **Figure 4**

The two graphs in Figure 4 appear to be the same (even though they aren't identical), indicating that we have an excellent fit. Based on how well $y=-\frac{1}{x^2}$ fits this data, our prediction for f'(x) is $y=-\frac{1}{x^2}$. This prediction is right on target! By definition,

$$f'(x) = \lim_{h \to 0} \frac{\frac{1}{x+h} - \frac{1}{x-h}}{2h}$$

$$= \lim_{h \to 0} \frac{(x-h)-(x+h)}{2h(x-h)(x+h)}$$

$$= \lim_{h \to 0} \frac{-2h}{2h(x-h)(x+h)}$$

$$= \lim_{h \to 0} \frac{-1}{x^2-h^2}$$

$$= -\frac{1}{x^2}$$

You will not be expected to verify your predictions of the derivative of function in this project, but careful curve fitting of the symmetric difference quotient data for functions in this project should lead to correct predictions.

Procedures:
I.   For each of the following functions, graph the symmetric difference quotient with h=.001, and from this graph predict a name for f'(x). Check your prediction for f'(x) by comparing its graph with the graph of the symmetric difference quotient. If your prediction appears to be "off," make the necessary adjustments and graphically check again. Repeat this guess and check procedure until you feel that you have correctly predicted f'(x).

	Function f(x)	Prediction for f'(x)
1.	f(x) = cosx	
2.	f(x) = cos(2x)	
3.	f(x) = cos(.5x)	
4.	f(x) = 4cosx	
5.	f(x) = 3 cos(2x)	
6.	f(x) = sinx	
7.	f(x) = sin(2x)	
8.	f(x) = sin(4x)	
9.	f(x) = sin(-2x)	
10.	f(x) = 3 sinx	
11.	f(x) = 1.5 sin(2x)	

12.   Look for patterns between the above functions and their derivatives and write a formula for f'(x) when f(x)=acos(bx). Do the same for f(x)= asin(bx).

13.   Does f(x)= x sin(2x) fit the above formula?   Use the symmetric difference quotient to help you decide and explain your answer.

II.   Set up a friendly window (see Appendix 2.5) centered around 0 with $\Delta x=.1$ and follow the directions in part II to predict f'(x) for each of the following functions.   Record your predicted function for each one.

	Function f(x)	Prediction for f'(x)
1.	f(x) = e^x	

2.   f(x) = e^{2x}
[Hint: compare the graph of the symmetric
         difference quotient with y=e^{2x}]

3.   f(x) = e^{3x}

4.   f(x) = $e^{.5x}$

5.   f(x) = e^{-x}
[Hint: compare the graph of the symmetric
         difference quotient with y=e^{-x}]

6.   f(x) = $4e^x$

Function f(x)   Prediction for f'(x)
7.   f(x) = 5e^x

8.   f(x) = 1/e^{2x}

9.   f(x) = 3e^{2x}

10.   f(x) = 2e^{4x}

11.   Write a formula for f'(x) where f(x)=be^{ax}.

12.   Does this formula apply to all 10 functions above? If not, list the ones for which the formula does not apply and tell why.

13.   Does this formula apply to f(x)=xe^x? Tell why or why not.

14.   Does this formula apply to f(x)=3e^4. Tell why or why not.

III. For each of the following functions, choose a candidate for the derivative of the function. Check your choice with the symmetric difference quotient with h=.001 over the interval $-\pi \le x \le \pi$.

1. $f(x) = \cos(x^2)$
      a) $f'(x) = -\sin(x^2)$ b) $f'(x) = 2\sin x$
      c) $f'(x) = -2x\sin(x^2)$ d) $f'(x) = -\sin(2x)$

2. $f(x) = \sin^2 x$
      a) $f'(x) = 2\sin x\cos x$ b) $f'(x) = \cos^2 x$
      c) $f'(x) = 2\sin x$ d) $f'(x) = 2\cos x$

3. $f(x) = e^{x^2}$
      a) $f'(x) = e^{x^2}$ b) $f'(x) = e^{2x}$
      c) $f'(x) = 2e^{x^2}$ d) $f'(x) = (2x)e^{x^2}$

4. $f(x) = e^{\sin x}$
      a) $f'(x) = e^{\sin x}$ b) $f'(x) = (\cos x)e^{\sin x}$
      c) $f'(x) = e^{\cos x}$ d) $f'(x) = (\cos x)e^x$

5. $f(x) = \sin(e^x)$
      a) $f'(x) = \cos(e^x)$ b) $f'(x) = e^x\cos(e^x)$
      c) $f'(x) = (\cos x)e^x$ d) $f'(x) = (\sin x)e^x$

6.   Consider the composition h of functions f and g denoted by h(x)=f(g(x)). Use the results of parts II, III and IV to find a general formula for h'(x) using the symbols f, g, f', g', and x.

      h'(x) =

Objectives:
1.   Given a function f, explore the existence of a derivative graphically, algebraically, and numerically.
2.   Develop visual images of the graph of a function which has no derivative at the point (a, f(a)).

Technology:
    TI-82, TI-85, or TI-92 graphing calculator.

Prerequisites:
1.   Definition of the derivative and right and left hand derivatives.
2.   Interpretation of a derivative as the slope of a tangent line and the limit of the slopes of secant lines.

Overview: The primary purpose of this project is to graphically explore whether a function y=f(x) has a tangent line at the point (a, f(a)). If a function has a tangent line at (a, f(a)) and the slope of this tangent is defined, then f is differentiable at (a, f(a)). Imagine magnifying a portion of the graph at (a, f(a)). If the magnified part of the curve looks linear, then this linear piece looks like its tangent line. Highly magnified portions of the graph of y=f(x) at a, b and c are provided in Figures 1, 2 and 3. The magnified portion looks linear in Figures 1 and 2, but it looks like the intersection of two rays in Figure 3. This evidence leads to the conjecture that a tangent line exists at a and b but not at c, in which case f would not be differentiable at (c, f(c)). Figure 2 indicates that the tangent line at (b, f(b)) may be vertical. If this is the case, the slope of the tangent line is undefined and f'(b) does not exist.

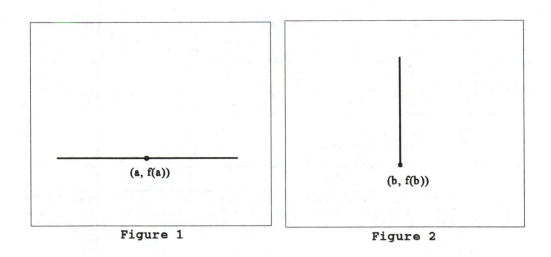

Figure 1   Figure 2

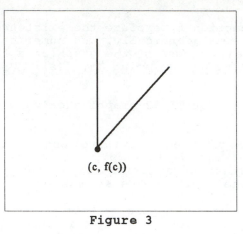

(c, f(c))

Figure 3

Procedures:
I.   Set up a friendly window (see Appendix 2.5) centered around
x=0 with Δx=.1 and set the zoom factors for x and y at 10 (see
Appendix 3.1). Choose appropriate values for ymin and ymax to
display the graph and choose zoom square so that the graph will
not look distorted. For each function f and each number a in
Table 1, use the zoom in feature (see Appendix 3.2) to magnify a
portion of the graph as much as possible. Based on the magnified
portion, make a conjecture about whether a tangent line exists at
(a, f(a)). If you think that a tangent line exists, determine
whether its slope is positive, negative, zero or undefined. As
you explore each problem, fill in Table 1. For details on
graphing piecewise functions, see Appendix 1.2.

$f(x)$	a	Tangent Line yes - no	Slope pos, neg, zero, or undefined	$f'(a)$ exists yes - no
1. $f(x) = \begin{cases} -4x^2+1, & x \geq 1 \\ -8x+5, & x < 1 \end{cases}$	$a=1$			
2. $f(x) = \begin{cases} x^2-3, & x \leq 2 \\ 3x-5, & x > 2 \end{cases}$	$a=2$			
3. $f(x) = 40x^3 - x^2$	$a=4$			

Table 1

f(x)	a	Tangent Line yes - no	Slope pos, neg, zero, or undefined	f'(a) exists yes - no
4. $f(x) = (x-1)^{\frac{2}{3}}$ $*^1$ see footnote	a=1			
5. $f(x) = \begin{cases} -2x^2 & , x \le 1 \\ -3.7x+1.7 & , x > 1 \end{cases}$	a=1			
6. $f(x) = \begin{cases} 5x^2-2, x \le 1 \\ 3x \quad , x > 1 \end{cases}$	a=.9			
7. $f(x) = |x-3|$	a=3			
8. $f(x) = \begin{cases} 4-x^2, x < 1 \\ 2+x^2, x \ge 1 \end{cases}$	a=1			

Table 1 (cont.)

II. Your calculator has a built-in numerical routine for estimating the value of the derivative of a function at a number, nDeriv on the TI-82 or TI-92 and nder on the TI-85 (see Appendix 8.1). For each function in Table 2 estimate f'(a) by using the appropriate feature on your calculator and fill in the second column of the table.

f(x)	a	Numerical estimate of f'(a)	Results from I: no tan line, vertical tan line, or non-vertical tan line
1. $f(x) = \begin{cases} -4x^2+1, x \ge 1 \\ -8x+5 \quad , x < 1 \end{cases}$	a=1		
2. $f(x) = \begin{cases} x^2-3, x \le 2 \\ 3x-5, x > 2 \end{cases}$	a=2		
3. $f(x) = 40x^3-x^2$	a=4		

Table 2

---

[1] Enter $(x-1)^{\frac{2}{3}}$ as $((x-1)^2)^{\wedge}(1/3)$

f(x)	a	Numerical estimate of f'(a)	Results from I: no tan line, vertical tan line, or non-vertical tan line
4. $f(x) = (x-1)^{\frac{2}{3}}$	a=1		
5. $f(x) = \begin{cases} -2x^2 &, x \le 1 \\ -3.7x+1.7 &, x>1 \end{cases}$	a=1		
6. $f(x) = \begin{cases} 5x^2-2, x \le 1 \\ 3x \quad\;, x>1 \end{cases}$	a=.9		
7. $f(x) = |x-3|$	a=3		
8. $f(x) = \begin{cases} 4-x^2, x<1 \\ 2+x^2, x \ge 1 \end{cases}$	a=1		

Table 2 (cont.)

III. The existence of a derivative at a number has been investigated graphically and numerically in Procedures I and II. Fill in the last column of Table 2 with your results from Procedure I if you haven't done so.

A. Determine in each case if the numerical estimate in column 2 does or does not support the graphical results in column 3. List the functions for which there is a discrepancy.

Table 2 above should indicate that your results from Procedure I do not always agree with those from Procedure II. It is possible that your graphical interpretation in Procedure I was incorrect. It is also possible that the numerical approximations from a calculator are misleading or incorrect.

B. For each of the functions whose numerical results do not agree with the graphical results, use the definition of the derivative to resolve the conflict by determining if f'(a) exists. Your investigation should involve checking out the left and right hand derivatives. Show your work and explain your final decision.

IV.
A.   Based on the results in Procedure III, did you find that the graphing techniques were sometimes misleading? If so, write an explanation of how one can be misled by using graphing techniques to determine if a function is differentiable at a number.

B.   Did you find that the numerical approximations of the derivative were sometimes misleading or incorrect? If so, write an explanation of how these numerical results may be misleading or incorrect.

Objectives:

1. Set up parametric equations that describe the motion of an object in one and two dimensions.
2. Use a function grapher to simulate motion.
3. Analyze graphs to determine where an object in motion is at time t.
4. Analyze the graph of the velocity function to determine how fast and in what direction an object is moving.

Technology:

TI-82, TI-85, or TI-92 graphing calculator.

Prerequisites:

1. A knowledge of first and second derivatives of polynomial functions.
2. An introduction to rectilinear motion.

Overview: In a two dimensional world, the position of an object in motion can be given by describing its position in both the x and y directions as a function of time. This can be described in symbols by $P_x=f(t)$ and $P_y=g(t)$. The parameter t represents time in seconds in this project. The parametric equations $P_x=f(t)$ and $P_y=g(t)$ determine the position of the object at any time t from a reference point called the origin. The velocity in the x and y direction is $v_x=f'(t)$ and $v_y=g'(t)$, respectively. v_x and v_y provide information about how fast and in what direction the object is moving. $a_x=f''(t)$ and $a_y=g''(t)$ give the acceleration in the x and y directions. If we assume that the acceleration in both the x and y directions is constant, then the position of the object can be described by:

$$(1) \qquad P_x=\frac{1}{2}a_x t^2+v_{x_0} t+P_{x_0}$$

$$(2) \qquad P_y=\frac{1}{2}a_y t^2+v_{y_0} t+P_{y_0}$$

a_x is the acceleration in the x direction, v_x is the initial velocity in the x direction, and p_{x_0} is the initial distance from the origin in the x direction. The parameters a_y, v_{y_0}, and p_{y_0} are defined in a similar manner.

Procedures:

I. Tossing a ball directly upward will result in motion which is one dimensional, or rectilinear motion. Suppose a ball is tossed upward with an initial velocity of 39.6 ft/sec from a position 31 feet above ground level. Because there are no forces in the x direction and hence no motion in the x direction, $P_x=0$. $P_y=-16t^2+39.6t+31$ since the acceleration due to gravity is $-32\,\dfrac{ft}{sec^2}$. Set your calculator in the degree mode and the parametric mode with a window of:

```
t =  0 to t = 3.1 and t-step = .1
x = -2 to x = 4    and x-scl  = 1
y = -1 to y = 60   and y-scl  = 10
```

Enter $x_{1T}=0$ and $y_{1T}=-16t^2+39t+31$ on the y= menu. See Appendix

1.3 for details about parametric graphing. Prior to graphing, turn off the axes (see Appendix 2.7) so that you can see the motion. Display the graph and use the trace to answer the following questions. Some of these questions cannot be answered by only looking at the graph of position as a function of time.

1.   How high does the ball go?

2.   How long does it take to reach its maximum height?

3.   How long is the ball in the air?

4.   What is its speed at the instant it hits the ground?

5.   Approximate the time or times when the ball is 40 feet above the ground.

6.   What is the speed of the ball at each time when the ball is 40 feet above the ground?

7.   What is the acceleration of the ball at t = 1 sec?

II.   The purpose of this procedure is to analyze two dimensional motion. A football is kicked at an angle of 26° with an initial velocity of v_0=75 ft/sec. The place where the ball is positioned just prior to kicking is the origin. The acceleration due to gravity is the only force acting on the football, if we ignore the drag force due to air resistance. Hence $a_y = -32 \dfrac{ft}{sec^2} = -\dfrac{32}{3} \dfrac{yds}{sec^2}$, and a_x=0.

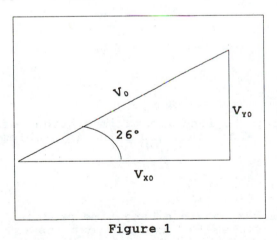

Figure 1

We will break the velocity into two components, an x-component, v_x, and a y-component, v_y. Since the magnitude of v_0 is 75 $\dfrac{ft}{sec}$ or $\dfrac{75}{3} \dfrac{yds}{sec}$ then $v_{x_0} = \dfrac{75}{3} \cos 26°$ and $v_{y_0} = \dfrac{75}{3} \sin 26°$. See Figure 1. From equations (1) and (2), the position of the ball is described by

$$P_x = 0t^2 + (\frac{75}{3} \cos 26°)t + 0 \quad \text{and} \quad P_y = -\frac{16}{3}t^2 + (\frac{75}{3} \sin 26°)t + 0.$$

Choose the parametric and the degree modes and enter the

formulas for P_x and P_y on the y= menu. Find an appropriate window that shows the motion of the football and use the tracer to answer the following questions.

1.   Explain the meaning of t, x, and y which appear when you trace the graph.

2.   How high does this kick go?

3.   How far does the football travel from where it is kicked to where it hits the ground?

4.   What is the hang time of this kick?

5.   If the initial speed is fixed, use the trace to find the kicking angle that produces a kick which yields the greatest distance from where it is kicked to where it hits the ground. (Hint: To investigate this question, change to $p_x = \left(\dfrac{75}{3} \cos B\right)t$ and $p_y = -\dfrac{16}{3}t^2 + \left(\dfrac{75}{3} \sin B\right)t$. Then the angle can be changed at the home screen by storing different values for the angle B. For example 10 [STO▸] B.) Record the angle and the distance the ball will travel before hitting the ground.

6.   Use algebra and trigonometry to find the angle which produces the maximum distance. Show your work.

7.   Use calculus to find the maximum height of the punt with the kicking angle that you found in problem 6. Show your work.

8.   Is the maximum possible hang time produced by the kicking angle in problem 6? Explain your answer. If not, what kicking angle will produce the maximum hang time?

9.   Use the trace to find the maximum height of such a kick.

Objectives:
1.   Use parametric equations to model the motion of a bullet and a ball.
2.   Simulate the motions on a calculator and determine if the bullet hits the ball.
3.   Use algebra to verify your conjecture.

Technology:
      TI-82, TI-85, or TI-92 graphing calculator.

Prerequisites:
1.   Introduction to rectilinear motion.
2.   Introduction to parametric equations.

Overview: A golf ball is suspended two meters above ground level. A toy gun aimed directly at the ball is mounted on a tripod 1.1m above the ground (see Figure 1).

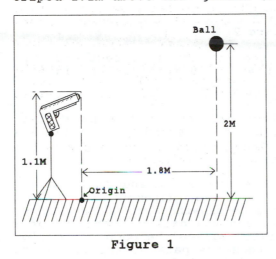

Figure 1

The point on the floor directly below the end of the barrel is designated as the origin. The point on the floor directly below the ball is 1.8m from the origin. The muzzle velocity of the gun is 10.5 m/s. The ball is released at the same instant that the bullet is discharged. We will determine whether the bullet will hit the ball. In two dimensional motion, the position of an object at any time t is given by describing its position in both the x and y directions as a function of time. If we assume that air resistance is negligible, then

$$P_x = \frac{1}{2} a_x t^2 + v_{x_0} t + P_{x_0}$$

$$P_y = \frac{1}{2} a_y t^2 + v_{y_0} t + p_{y_0}$$

The parameter t in each function represents time in seconds. The parameters a_x and a_y represent the x and y components of acceleration. The initial velocities in the x and y directions are represented by v_{x_0} and v_{y_0} and the initial distances are represented by p_{x_0} and p_{y_0}. Since the motion of the ball is strictly vertical, the x components of acceleration and velocity are both 0, so p_x is a constant.

Procedures:
1.   Find and record the parametric equations which describe the motion of the ball (recall that $a_y = 9.8$ m/s^2). Set your calculator in the parametric mode and enter these equations at the y= menu. See Appendix 1.3 for details about parametric graphing.

The motion of the bullet is more complex since it is not rectilinear motion.

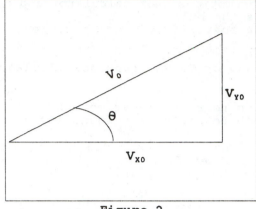

Figure 2

The initial velocities in the x and y directions are $v_{x_0}=10.5\cos\theta$ and $v_{y_0}=10.5\sin\theta$, since 10.5 m/sec is the initial velocity of the bullet. See Figure 2.

2. Use the above information to set up the parametric equations for the motion of the bullet. Record these equations here and enter them on the y= menu.

Change the mode to dot rather than connected and simultaneous rather than sequential. Set the time increment (tstep) at $\Delta t=.05$. This means that you will be showing a video "snapshot" of the position of the bullet and the ball every $\frac{1}{20}$th of a second. Adjust the window so that you see the complete motion of the ball and the bullet. Trace the path of the bullet to the point closest to the path of the ball and complete the first row of Table 1. Without touching the left or right arrow keys, switch the cursor to the path of the ball and fill in the second row of Table 1.

	t	x	y
Bullet			
Ball			

Table 1

3. Interpret the meaning of t, x, and y for the bullet and for the ball.

4. Do you think that the bullet hits the ball? Tell why.

Next, change your window settings to focus on the point where the collision would take place, set Δt=.01, and repeat the above directions to fill in Table 2.

	t	x	y
Bullet			
Ball			

Table 2

5.   **G1** Print (see Appendix 4.1) the graph showing the motion of the bullet and the ball for Δt=.01. Now do you think that the bullet hits the ball? Tell why.

6.   Use algebra to verify your conclusions. Show all work in an organized manner.

Checklist of calculator graph printouts to be handed in:
☐   **G1**   Parametric graph of the position of the bullet and the ball for Δt=.01.

Objectives:
1.   Use a CBL to collect distance and force data for motion created by hanging a mass on the end of a spring.
2.   Find a model which describes the displacement of the mass at any time t and use it to derive a model for acceleration.
3.   Analyze experimental data to determine if the data is consistent with Newton's Second Law of motion and Hooke's Law.

Technology:
     TI-82, TI-85 (CBL compatible version), or TI-92 graphing calculator, CBL, motion detector, force sensor, mass, and programs HARMONIC, VIEWS85, TRACE85.

Equipment:
1.   A spring whose length is between 30cm and 80cm.
2.   A primary mass of about 4 times the mass of the spring. (The mass should be about 200g to 600g.)
3.   A rod and a support.   (See Figure 2).
4.   Three other masses between 100g and 600g.

Prerequisites:
1.   Given a set of points which appear to lie on the graph of a function from the family $s = a \cos(\omega t)$, find the parameters a and ω which define the particular function.
2.   Find $\dfrac{d^2s}{dt^2}$ for a function defined by $s = a \cos(\omega t)$.

Overview of Procedure I: A mass is attached to the end of a spring, pulled down below the equilibrium position and released. For each time interval, the distance in meters from a motion detector and the spring force in Newtons is recorded by a CBL and stored in a calculator.

Procedures:
I.   Each team member is to collect the data described in Procedure I.   For example, a team of three should have three different sets of data.

1.   Connect a CBL to your calculator with a unit-to-unit link cable.   Connect a motion detector to the sonic port of the CBL, and use a CBL DIN adapter to attach a force sensor to channel 1 of the CBL.

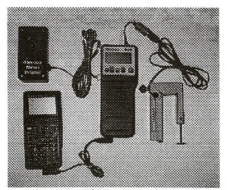

Figure 1

2.   Connect the force sensor to a rod and attach a spring to the force sensor. Attach your primary mass securely to the end of the spring and position the motion detector on the floor facing upward as shown in Figure 2. The mass should be between .7m and 1.2 m from the motion detector.

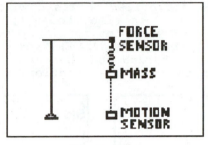

Figure 2

Let the mass hang so there is no spring motion. This is called the equilibrium position and our reference system will be defined so that the displacement, s, is zero at this point. The choice of a reference system is arbitrary. Because of the position of the motion detector, we will define s so that when the mass is below the equilibrium position, the directed displacement s is negative and when it is above, s is positive. See Figure 3.

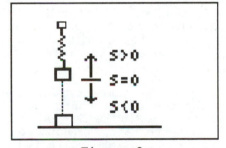

Figure 3

3.   Turn on the CBL and run program Harmonic (Please note: to run the HARMONIC program on a TI-85, you must have VIEWS85 and TRACE85 in your calculator). From the main menu (which is displayed in Figure 4), choose option 1 to collect distance, force, and time data. When prompted, enter 20 for the

Figure 4

number of samples per second. At the prompt "Press Enter to Start," have one team member pull the mass below equilibrium a distance which is approximately equal to one fourth the natural

length of the spring and hold it there. The mass should be at
least a half meter from the floor. Press ENTER and release the
mass shortly after without pulling or pushing it.   It should
oscillate in a vertical path above the motion detector without
swaying.   The data is collected over a time interval of about 5
seconds.   When the data collection is completed, the main menu
will be displayed.   At this time choose option 2 and then the
"DIST-ONLY" option from the plot menu.   (Please note: to return
to the menu at any time, press ENTER)   Your graph should look
linear for a short time period and then have the characteristics
of the graph of a sine or cosine function.   (See Figure 5.)

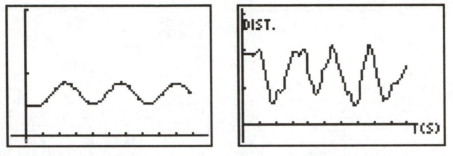

Figure 5   **Figure 6**

If your graph looks "jerky" like the one in Figure 6 you probably
pulled the mass down too far. It is also possible that the mass
was "swaying" rather than oscillating in a vertical path.   In
either case repeat the data collection as described above.
Continue to collect data until each team member has a sample of
data points whose distance plot is similar to the one in Figure
5.

4.   At the beginning of the time period, the distance from the
mass to the motion detector is not changing.   This corresponds to
the horizontal segment in Figure 5.   At these times the data
collection has begun but the mass has not been released. We will
delete all of the points which were collected before the mass was
released, except for the last one.   In other words, the linear
piece of the curve will be deleted, leaving a cosine curve.
Please note that slight variations in the distance values may
occur before the mass was released, due to "noise" in the
measuring system.   Trace (you are already in the trace mode) the
scatter plot of distance vs. time to determine the number of
points to delete.   When you decide on the number of points to
delete, return to the main menu, choose option 4, and when
prompted, input the number of points to be deleted and press
ENTER.   Choose option 3 to graph the distance vs. time for the
adjusted data.   If you think that you have not deleted enough or
possibly too many points you can repeat this part (#5) again.
The same number of points that are deleted from the distance data
are also deleted from the force data.   Record the number of
points that you have deleted from the original data.

5.   The purpose of this part is to find a function which gives
the displacement of the mass from the equilibrium position at any
time t.   The graph of the adjusted distance vs. time should now
look similar to the graph in Figure 7.   These distances are not
the displacement from equilibrium but the distance of the mass
from the motion detector.   The graph of displacement vs. time is
a vertical translation of the graph in Figure 7.   The translated
graph will have the equilibrium position on the x-axis, as shown

in Figure 8.

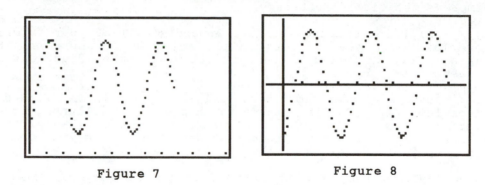

Figure 7   **Figure 8**

Select the TRANSLATE option from the main menu. Read the directions and trace to help you estimate the amount of the translation and press ENTER (F5 on a TI-85). Now type in the directed amount of the translation and press ENTER. You will be given the option to trace and make any necessary adjustments to the translation. If you have made the proper translation, the extreme downward displacement from equilibrium (s=0) should be the opposite of the extreme upward displacement. If you are off by more than .03 meters, adjust your translation. You will now be shifting the graph that you see. Record the directed net amount of your translation. Press ENTER and choose 1:YES if you want to translate more or 2:NO to return to the main menu.

original time	List 1
original distance	List 2
original force	List 3
adjusted time	List 4
adjusted distance	List 5
adjusted force	List 6

Table 1

6.   Each team member should be analyzing his or her own data and each is to complete analysis sheet I.

Analysis sheet for Procedure I
1.   Draw a scatter plot of the adjusted distance vs. time data. The storage lists for the data are provided in Table 1. See Appendix 1.5 for information about graphing scatter plots. This graph appears to be the graph of a function which belongs to the family of functions described by s=a cos(ωt). Trace the curve to find the amplitude and record your result. (Use the free-moving cursor on a TI-85).

2.   Record the period for your graph.

3.   Use the period and amplitude to record a function which models your curve in the form s=a cos(ωt). You should compare the graph of this function with the scatter plot of adjusted

distance vs. time and you should make adjustments to a and ω to improve the fit if necessary.

4.   **G1** Print out a graph (see Appendix 4.1) of your function overlaid on a scatter plot of the adjusted distance vs. time data. If your function does not fit well, make adjustments in question 1-3 above.

5.   Draw a scatter plot of the original distance vs. time data (see Table 1). To fit a curve to this data we will need both a vertical translation of the function in question 4 above and a phase shift. Record the number of data points that you deleted in step 5 of Procedure I.

6.   Using the fact that the data points were obtained in intervals of .05 seconds, find the amount of the phase shift.

7.   Record the amount of the vertical translation from step 6 of Procedure I.

8.   Use the phase shift and translations in questions 6 and 7 to record a function of the form s=a cos(ωt+c)+d which models the original distance vs. time data.

9.   **G2** Print out a graph of your function overlaid on a scatter plot of the original distance vs. time data. If your function does not fit well, make adjustment to questions 5-8 above.

As a team, choose one team member's experimental data to be analyzed in Procedure II. The data which has the best fitting function (G1 question 4) should be chosen for Procedure II.

Overview of Procedure II: Newton's Second Law, Force = Mass × Acceleration, and Hooke's Law, Force = κs are central to the theory of simple harmonic motion. Keep in mind that one seeks a motion function, s= f(t), where s is the directed displacement of the object from the equilibrium position at time t.   Since $a(t) = \dfrac{dv}{dt} = \dfrac{d^2s}{dt^2}$, Newton's Second Law can be written as $F = m\dfrac{d^2s}{dt^2}$.

The model for simple harmonic motion assumes that there are no external forces acting on the mass. This assumption is not completely correct since the air will offer some resistance to motion of the mass. However, since the viscosity of the air is low, the effect of this viscous drag will be minimal. If the spring were oscillating in a cylinder of oil, however, there would be a significant viscous drag force acting on the mass. This force would retard the motion of the spring.

Analysis sheet I should have been completed by each team member. The conclusion in Procedure II depends on how well the function of the form s= a cos(ωt), found in Procedure I, fits the adjusted distance vs. time data. For this part the team should have selected the data and the corresponding function which best fits the data in Procedure I. Once you transfer the "best" data to your calculator for Procedure II, your data in part I will be lost, so complete analysis sheet I prior to proceeding with Procedure II.

II.
1.   **G3** Print a scatter plot of force (L6-ylist) vs. distance (L5-xlist).

2.   Does the shape of this scatter plot support Hooke's Law, F=ks? Explain.

3.   Fit a curve to this data and use this information to estimate the spring constant k. Record the equation of the curve. (See Appendix 7.2, 7.3, 7.4, or 7.5 for curve fitting instructions, depending on what type of relationship you expect).

4.   Select 4 masses, hang each mass on the end of your spring and measure the displacement from the natural length of the spring for each one. Fill in Table 2 and use the correct units.

Mass (kg)	Force, F (N)	Displacement, s (m)

Table 2

5.   Use the information to estimate the spring constant k. This result should be close to |k| in question 3.

6.   Use the function $s=a\cos(\omega t)$ from Procedure I and the experimental force data to show that the experimental data is consistent with Newton's Second Law, F=ma. **G4** Include printouts of any graphs which may help your explanation.

Checklist of calculator graph printouts to be handed in:
☐   **G1**   Print a graph of your function overlaid on a scatter plot of adjusted distance vs. time.
☐   **G2**   Print a graph of your function overlaid on a scatter plot of the original distance vs. time data.
☐   **G3**   Print a scatter plot of force vs. distance.
☐   **G4**   Print a scatter plot and/or graphs to support your explanation.

Objectives:
1.   Develop an understanding of how the derivative can be used in various situations.
2.   Develop an understanding of the graphical interpretations of the derivative.

Technology:
     None required.

Prerequisites:
1.   Basic introduction to the derivative as a rate of change, and as the slope of the tangent line to a curve at a point.
2.   Introduction to the second derivative as a measure of the concavity of a function.

Overview: The symbols $\frac{dy}{dx}$ can be interpreted as the rate of change in y with respect to x. For example, if P represents the population of Iceland as a function of time, t (in years), then $\frac{dP}{dt}$ represents the rate of change in the population of Iceland with respect to time. Suppose that $\frac{dP}{dt}$=27,000. Then the population of Iceland is currently increasing at a rate of 27,000 people per year. If $\frac{dP}{dt}$ was -5,000, then the population would be decreasing at the rate of 5,000 people per year.

Procedures:
I.   Use the interpretation of the derivative as a rate of change to work through each application.

A.   Economists define an individual's "marginal propensity to consume" as follows:

$$\text{Marginal propensity to consume}=\frac{dC}{dI},$$

where C is the consumption function - the amount that an individual consumes (or spends), and I is the individual's income.

1.   What does it mean if $\frac{dC}{dI}$=1?

2.   What does it mean if $\frac{dC}{dI}$=0?

3.   Why would it be unnatural to have $\frac{dC}{dI}$<0?

4.   How would you explain the meaning of $\frac{dC}{dI}$ to someone who does not know calculus?

5.   Explain in words what it means if $\frac{dI}{dt}$<0, where t is time.

6. Suppose that $\frac{dI}{dt}<0$ and $\frac{dC}{dI}>0$. Explain why $\frac{dC}{dt}<0$.

7. How would you explain the meaning of $\frac{dC}{dt}$ to someone who does not know calculus?

B. Suppose that Alice and Betty are jogging on a straight path and that their distances from their starting point are a and b respectively (See Figure 1).

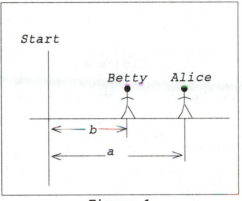

Figure 1

1. How would you explain the meaning of $\frac{da}{dt}$ to someone who does not know calculus?

2. Is it possible, as Figure 1 is drawn, for $\frac{db}{dt}>\frac{da}{dt}$ for some time t? If so, describe one possible scenario.

3. How would you explain the meaning of $\frac{da}{db}$ to someone who does not know calculus?

4. If $\frac{da}{dt}>\frac{db}{dt}>0$, what can you say about the sign of $\frac{da}{db}$?

5. Is it possible to have $\frac{da}{db}<0$? If so, describe one possible scenario.

C. Let V represent the volume of water in the cylindrical tank shown in Figure 2. A valve is opened at the bottom of the tank and water is allowed to flow out of the tank for a short period of time before the valve is closed.

1.   How would you explain the meaning of $\frac{dV}{dt}$ to someone who does not know calculus?

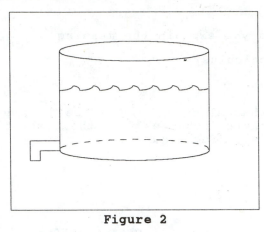

Figure 2

2.   Draw a rough sketch of $\frac{dV}{dt}$ vs. time in Figure 3 staring at a point t_1 before the valve is opened, continuing through the time when the valve is opened, t_2, and closed, t_3, to a point t_4 after the valve is closed.

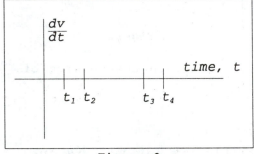

Figure 3

II.   Suppose that Fred, Gilda, and Hersey are all jogging on the same straight path and that their distances from a reference point on that path are given as functions of time F(t), G(t), and H(t), respectively. See Figure 4.

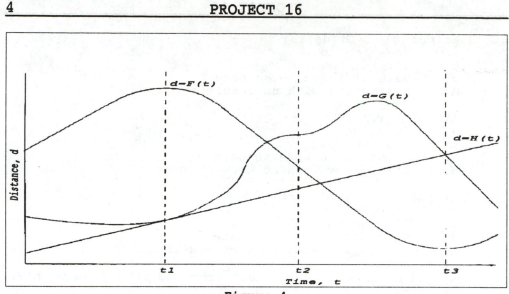

Figure 4

A.   Based on the information in Figure 4, insert the correct relationship >,<, or = for each of the following.

1.   For time t_1:

    F(t)____G(t)   F'(t)____G'(t)   F"(t)____G"(t)
    G(t)____H(t)   G'(t)____H'(t)   G"(t)____H"(t)
    F(t)____H(t)   F'(t)____H'(t)   F"(t)____H"(t)

2.   For time t_2:

    F(t)____G(t)   F'(t)____G'(t)   F"(t)____G"(t)
    G(t)____H(t)   G'(t)____H'(t)   G"(t)____H"(t)
    F(t)____H(t)   F'(t)____H'(t)   F"(t)____H"(t)

3.   For time t_3:

    F(t)____G(t)   F'(t)____G'(t)   F"(t)____G"(t)
    G(t)____H(t)   G'(t)____H'(t)   G"(t)____H"(t)
    F(t)____H(t)   F'(t)____H'(t)   F"(t)____H"(t)

B.   Choose the correct response or responses for each statement.

1.   At point t_1, Gilda is:
    a.   Jogging toward the reference point.
    b.   Jogging away from the reference point.
    c.   Stationary.

2.   At point t_2, Gilda is:
    a.   Jogging toward the reference point.
    b.   Jogging away from the reference point.
    c.   Stationary.

3.   At point t_3, Gilda is:
    a.   Jogging toward the reference point.
    b.   Jogging away from the reference point.
    c.   Stationary.

4.   At point t_1, Gilda is:
      a.   Speeding up.
      b.   Slowing down.
      c.   Maintaining an even pace.
      d.   Stationary with no acceleration.

5.   At point t_2, Gilda is:
      a.   Speeding up.
      b.   Slowing down.
      c.   Maintaining an even pace.
      d.   Stationary with no acceleration.

6.   At point t_3, Gilda is:
      a.   Speeding up.
      b.   Slowing down.
      c.   Maintaining an even pace.
      d.   Stationary with no acceleration.

7.   At point t_1, who is furthest from the reference point?
      a.   Fred
      b.   Gilda
      c.   Hersey

8.   At point t_2, who is furthest from the reference point?
      a.   Fred
      b.   Gilda
      c.   Hersey

9.   At point t_3, who is furthest from the reference point?
      a.   Fred
      b.   Gilda
      c.   Hersey

10.   At point t_3, the person who is jogging the fastest is:
      a.   Fred
      b.   Gilda
      c.   Hersey

Objectives:

1.   Develop an intuitive feel for the definite integral.
2.   Introduce an application of the definite integral.
3.   Discuss the relationship between distance, velocity, and
      acceleration both graphically and numerically.

Technology:
      TI-82, TI-85, or TI-92 graphing calculator

Prerequisites:
      None

Overview[1]:   The art of directing a vehicle from one place to
another, called navigation, was first developed by the ancient
mariners.   The Vikings, for example, utilized the "Pilotage
Method" of navigation to determine the course and position of
their craft.   This technique used visual observation of prominent
land masses to steer the ship from point to point.

Somewhat later, dead-reckoning navigation came into extensive
use.   A log of direction and velocity was maintained and plotted
on a chart.   The resulting plot indicated the position of the
vessel.   The errors introduced by ocean currents and wind were
corrected by pilotage at the first sight of land.   As modes of
travel changed, man's quest for faster means of transportation
forced the "art" of navigation to give way to the "science" of
navigation.   The science of navigation may be defined as the
application of calculations to determine the position of a
vehicle and to direct it to a predetermined destination.   With
the advancement of radio techniques, positioning systems such as
LORAN, VOR, DME, and DOPPLER were developed to aid the navigator.
As air transportation reached transcontinental proportions, radio
positioning and celestial techniques were further developed to
satisfy the faster and more accurate navigational requirements.
However, radio systems are somewhat limited, as they require
extensive networks of ground stations and are subject to both
manmade and natural interference.   In the 1950's, the Department
of Defense recognized the need for a navigational system that did
not require a reference to its external environment.   The
Massachusetts Institute of Technology developed the first
aircraft Inertial Navigation System.   This system was completely
self-contained and required neither radio nor visual inputs to
determine position or direction.   The ability to navigate
completely without contact with the external environment of the
vehicle makes inertial navigation systems extremely desirable in
today's aviation industry.

In order to understand an inertial navigation system we must
consider both the definition of "inertial" and the basic laws of
motion as described by Newton over 300 years ago.

Inertia can be defined as follows: "A body continues in a state
of rest, or uniform motion in a straight line, unless acted upon
by an external force."   This is also known as Newton's first law
of motion.

---

[1] Primarily from <u>Avionics Past & Present</u> by Honeywell©

Newton's second law of motion states: "The acceleration of a body is directly proportional to the sum of the forces acting on the body."

Newton's third law of motion states: "For every action, there is an equal and opposite reaction."

The basic measuring instrument of the inertial navigation system is the accelerometer (see Figure 1.)

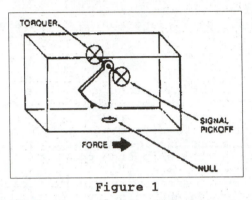

Figure 1

Two accelerometers are mounted in the system. One will measure the aircraft's acceleration in the North-South directions, and the other will measure the aircraft's acceleration in the East-West directions. The accelerometer is basically a pendulous device. When the aircraft accelerates, the pendulum, due to inertia, swings off its null position. A signal pick off device tells how far the pendulum is off the null position. The signal from the pick off device is sent to an amplifier, and current from the amplifier is sent back into the accelerometer to the torquer motor. The torquer motor will restore the pendulum back to its null position.

The acceleration signal from the amplifier is also stored in a computer list to produce a set of acceleration vs. time data. The acceleration information is then _integrated_ to produce velocity data which in turn is _integrated_ to produce distance data. This distance data is then used to determine the position of the plane at any time. The computer associated with the inertial system knows the latitude and the longitude of the takeoff point and calculates that the aircraft has traveled so far in a North-South direction and so far in an East-West direction. It now becomes simple for a digital computer to continuously compute the new present position of the aircraft.

This type of navigation system is also very useful on submarines.

The system also has mechanisms to compensate for tilt in the plane and rotation of the earth while the plane is in flight. The heart of the system, though, is the ability to collect acceleration data.   Inertial navigation depends on the _integration_ of acceleration to obtain velocity and distance. That is the subject of this project.

Figure 5
Distance
Feet

Figure 4
Velocity
Ft/Sec

Figure 3
Acceleration
Ft/Sec^2

Figure 2
Acceleration
Ft/Sec^2

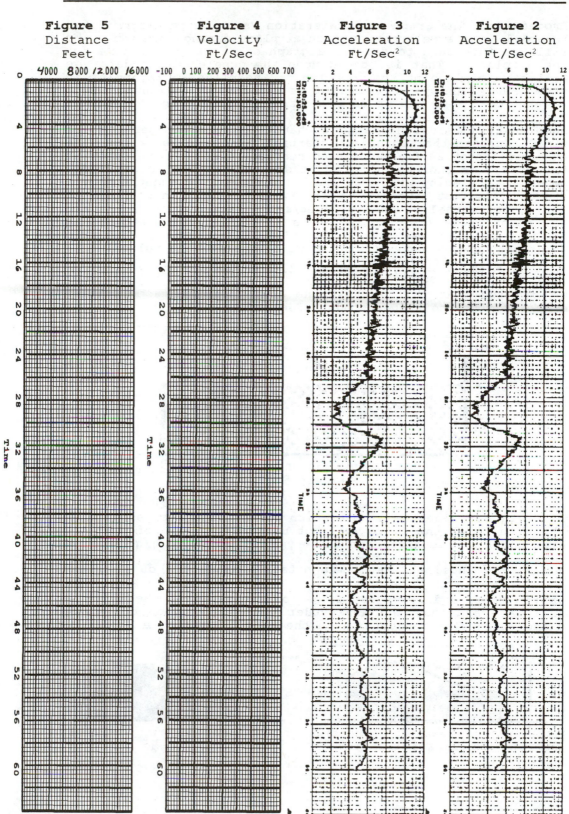

Procedures: The graph of acceleration vs. time in Figure 2 was recorded by an accelerometer on an airplane during take off.[2] We will use this data to develop a graph of distance vs. time for the airplane. That is, based upon the accelerometer data, we will try to pinpoint the location of the plane at any point in time. The plane starts from rest at a point on the runway which we will call x=0, and moves in a straight path down the runway.

Acceleration is defined as the change in velocity per unit of time. Suppose that an object starting from rest has a constant acceleration of 20 ft/sec^2 for 60 seconds. Then the velocity of the object is increasing by 20 ft/sec per second, so after one second the velocity will increase by 20 ft/sec. Since the object started from rest (i.e. Zero velocity initially) the velocity of the object at t=1 second is 20 ft/sec. Between t=1 second and t=2 seconds the velocity increases by another 20 ft/sec, so at t=2 seconds, the velocity is 40 ft/sec. The process continues for the entire 60 seconds so that at t=60 sec. the velocity is:

$$v = (20 \; \frac{ft/sec}{sec})(60 \; sec.) = 1200 \; ft/sec$$

We can see that this is just the area under the acceleration graph from t=0 to t=60 seconds. See Figure 6.

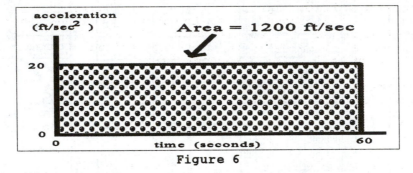

Figure 6

I.   In our case the acceleration is not constant, so finding the area under the curve will not be so easy. We refer to the area under the acceleration curve y=a(t), shown in Figure 7, as a definite integral, using the notation $Area=\int_a^b a(t)\,dt$. We can obtain upper and lower estimates for the value of the definite integral, and thus the area under the acceleration curve, by using rectangles to approximate the area. See Figure 8.

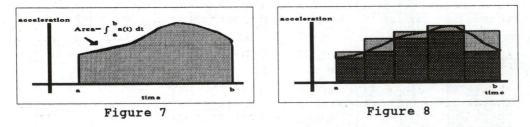

Figure 7   **Figure 8**

---

[2] Data from Gulfstream Aerospace Corp.

Adding the areas of all of the shorter rectangles will give us an underestimate of the actual area under the acceleration curve. Adding the areas of all of the taller rectangles will give an overestimate of the actual area.

Returning now to our airplane data in Figure 2, consider the graph of acceleration vs. time in the time interval from t=0 to t=10 seconds. We can see from the graph that in this time interval, the acceleration is always greater than 5.2 ft/sec^2 so an underestimate of the velocity after 10 seconds is:

$$\text{underestimate of velocity at t=10 sec} = (5.2\ \frac{ft/sec}{sec})(10\ \text{sec.}) = 52\ ft/sec.$$

Go to the velocity vs. time graph in Figure 4 and mark the point with coordinates t=10 sec and v=52 ft/sec. Shade in the area of the rectangle on the acceleration graph of Figure 2 from t=0 to t=10 with a height of 5.2 ft/sec. The area of this rectangle is an underestimate of the actual area under the curve from t=0 to t=10 seconds. This area corresponds to an underestimate of the velocity at t=10 seconds.

Next consider the portion of the graph of acceleration vs. time from t=10 to t=20 seconds. We can see from the graph that in this time interval, the acceleration is always greater than 6.4 ft/sec^2 so an underestimate for the amount of increase in velocity from t=10 to t=20 seconds is:

$$\text{underestimate of increase in velocity from t=10 to t=20 sec.} = (6.4\ \frac{ft/sec}{sec})(10\ \text{sec}) = 64 ft/sec$$

Since the velocity at t=10 seconds was estimated to be 52 ft/sec, we get

$$\text{underestimate of velocity at t= 20 sec.} = 52 + 64 = 106\ ft/sec.$$

Go to the velocity vs. time graph in Figure 4 and mark the point with coordinates t=20 sec and v=106 ft/sec. Shade in the area of the rectangle on the acceleration graph of Figure 2 from t=10 to t=20 with a height of 6.4 ft/sec. The area of this rectangle is an underestimate of the actual area under the curve from t=10 to t=20 seconds.

Continue this process to obtain an underestimate of the graph of velocity vs. time in Figure 4 from t=0 to t=60 seconds. Connect these points with line segments and label this curve "Δt=10". At this point you should also have 6 rectangles shaded in Figure 2.

II. Using the acceleration graph $\frac{ft/sec}{sec}$ in Figure 3, which is identical to the one given in Figure 2, repeat Procedure I, but this time obtain an overestimate of the velocity. For example, in the first 10 second interval, the acceleration is always less than 11.4 ft/sec^2. Hence, an overestimate of the velocity at t=10 seconds is:

$$\text{Overestimate} = (11.4\ )(10\ \text{sec}) = 114\ ft/sec$$

Plot the graph of your overestimates for velocity on the grid in Figure 4 and label it "Δt=10" also. Shade the area of the rectangles on the acceleration graph in Figure 3.

III.   Repeat procedures I and II but this time use a time interval of two seconds instead of 10 seconds. Plot both the overestimate and the underestimate on the same velocity axes as your previous plots (Figure 4) and label them both "Δt=2". Shade only the area corresponding to the first 5 rectangles (from t=0 to t=10 seconds) on the graphs of acceleration vs. time in Figure 2 and Figure 3. Use a different color pen or pencil than was used to shade the rectangles corresponding to Δt=10 sec.

1.   Relative to the two curves you've sketched on Figure 4, where is the actual graph of velocity vs. time located?

2.   Which graph will allow you to more accurately predict the shape and location of the actual velocity graph, the one obtained from Δt=10 seconds or the one obtained from Δt=2 seconds?

3.   Referring to the shaded rectangles in Figures 2 and 3, explain why you think that one graph in Figure 4 is better than the other?

4.   How could you improve your prediction of the shape and location of the actual velocity graph?

IV.   In Procedures I through III we used the acceleration graph to develop a prediction for the velocity graph. Using the same approach, we can use the velocity graph to develop a prediction for the distance vs. time graph.

Velocity is defined as the change in distance per unit time. Suppose an object starting at t=0 has a constant velocity of 30 ft/sec for 60 seconds. Then the distance traveled by the object is increasing by 30 ft/sec, so after one second, the distance traveled by the object will increase by 30 feet. Then at t=1 sec, d=30 feet. Between t=1 and t=2 seconds, the distance increases by another 30 feet, so at t=2 seconds, d=60 feet. The process continues for the entire 60 seconds, so the distance at t=60 seconds is: $d = (30 \frac{ft}{sec})(60 \ sec) = 1800$ feet

We can see that this is just the area under the velocity graph from t=0 to t=60 seconds. See Figure 9.

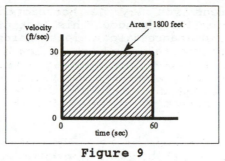

Figure 9

In our case the velocity is not constant but we can approximate the area under the velocity curve by using rectangles as we did earlier for the acceleration graph.

To obtain an underestimate of the distance curve we will use the same method that we used in Procedure I to obtain the underestimate of the velocity curve. Using the underestimate of the velocity curve (with Δt=2) in Figure 4, the velocity from t=0 to t=2 second is always greater than v=0 ft/sec, so an underestimate of the distance traveled from t=0 to t=2 second is:

     Underestimate of the   = (0 ft/sec)(2 sec) = 0 feet
     distance at t=2 sec.

Using the same method that was outlined in Procedure I, obtain an underestimate of the distance curve and plot it on Figure 5.

V.   In a similar way, use the overestimate of the velocity curve (with Δt=2) to obtain an overestimate of the distance graph. Plot it on the same graph as the underestimate in Figure 5.

1.   Relative to the two curves you've sketched in Figure 5, where is the actual graph of distance vs. time located?

Recall that our goal was to use acceleration information to determine the location of an airplane at any time. To do that we first need to <u>approximate</u> the velocity, and then use it to <u>approximate</u> the distance. Hence, the errors in approximating the velocity are magnified in the approximation of the distance.

2.   How can we reduce the error in our approximations, and thereby improve our approximation of distance?

3.   The accelerometer collects data at discrete points, with a small time interval between data readings (See Figure 10). The graph in Figure 2 was obtained by simply connecting such points with line segments. This is different from the types of

continuous functions you may be accustomed to studying in
calculus (see Figure 11). How does this lack of continuity limit
the accuracy of our distance calculations, which are based on the
discrete acceleration data?

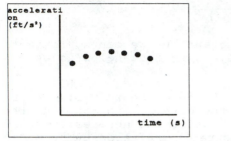

Figure 10

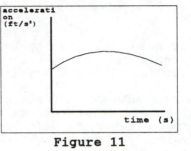

Figure 11

Objectives:
1. Approximate a definite integral by using right- or left-hand sums.
2. Explore the relative size of the value of a definite integral, a right-hand sum and a left-hand sum.
3. Devise a method for finding a bound for the error when left- and right-hand sums are used to approximate a definite integral.

Technology:
TI-82, TI-85 or TI-92 graphing calculator, program RLAREA

Prerequisites:
1. An introduction to Riemann sums and the definite integral.
2. An introduction to the function properties of increasing, decreasing, concave up and concave down.
3. An introduction to summation notation.

Overview: The main purpose of this project is to approximate a definite integral $\int_a^b f(x)\,dx$ and to explore ways of finding a bound for the error. There are several methods for approximating definite integrals. Left- and right-hand Riemann sums are used in this project. Even though there are methods that give more accurate results with less effort, right- and left-hand sums are used to provide a connection between Riemann sums and definite integrals. Under certain conditions it is easy to find a bound for the error when using left- and right-hand sums.

Example 1: The left-hand sum with n=4 for $\int_a^b f(x)\,dx$ is illustrated in

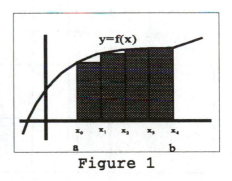

Figure 1

Figure 1. The sum of the areas of the four rectangles is called a left-hand sum with n=4. The name left-hand sum is used because the height of any rectangle is the function value of the left end point of the base of that rectangle. The interval from a to b is divided into four subintervals each of width $\Delta x = \frac{b-a}{4}$.

The area of the first rectangle is $f(x_0)\Delta x$, the area of the second rectangle is $f(x_1)\Delta x$, and so on. The sum of the areas of the four rectangles is represented by

$$\sum_{k=1}^{4} f(x_{k-1})\Delta x.$$

Example 2: The right-hand sum with n=4 for $\int_a^b f(x)\,dx$ is

illustrated in Figure 2. The name right-hand sum is given because the height of a given rectangle is the function value of the right end point of the base of that rectangle. The area of the first rectangle is $f(x_1)\Delta x$, the area of the second rectangle is $f(x_2)\Delta x$, and so on. The sum of the areas of the four rectangles is

$$\sum_{k=1}^{4} f(x_k)\Delta x.$$

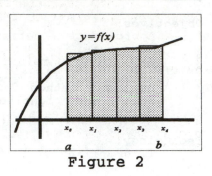

Figure 2

A program called RLAREA will be used to view and calculate left- and right-hand sums associated with a definite integral. Your job is to explore how to determine a maximum error when using left- or right-hand sums.

Procedures:
I.   For each integral in Table 1, store the integrand at y1, and view the graph on the interval [a, b]. Based on this graph, check the appropriate function characteristics on [a, b] - increasing, decreasing, concave up, and concave down - in Table 1. Then run the program RLAREA to compute the left- and right-hand sums. You will need to input the endpoints of the interval [a,b] and the number of subintervals (use 8 subintervals). Choose option 1 to show the graphs. Record the values for the left- and right-hand sums in Table 1. In the last column record the relationship between LHS, RHS, and the actual area under the curve, A. For example, LHS<A<RHS.

$A=\int_{a}^{b} f(x)\,dx$	INC	DEC	CUP	CDN	LHS	RHS	Write LHS, RHS and A in increasing order
$\int_{.5}^{3} e^{1/x}\,dx$							
$\int_{0}^{\pi/2} \cos x\,dx$							
$\int_{0}^{\pi/2} \sin x\,dx$							
$\int_{0}^{2} e^{x}\,dx$							

Table 1

Base your answers to the following questions on Table 1.

1.   Under what condition is the LHS larger than the value of the integral?

2.   Under what condition is the RHS larger than the value of the integral?

3.   Do the characteristics of increasing vs. decreasing determine the order of LHS, RHS and A?

4.   Do the characteristics of concave up vs. concave down determine the order of LHS, RHS and A?

5.   Write a rule to determine the relative size of RHS, LHS, and A, based on one of the function characteristics from #3 or #4 above.

6.   Draw sketches to support your conjectures for #5.

7.   Write a rule for finding a bound for the maximum error when using LHS and RHS to approximate $\int_{a}^{b} f(x)\, dx$.

II.

1.   Fill out Table 2 for the integral $\int_{.5}^{3} e^{1/x} dx$ by storing the function $y=e^{1/x}$ at y_1 and running the program RLAREA to calculate the RHS and the LHS.

n	value of LHS	value of RHS	maximum possible error
5			
10			
20			
40			
60			

Table 2

2.   Does your rule in #7 apply to this integral? Why?

3.   Describe the trend for the maximum possible error as n increases.

4.   Find the smallest value of n that produces an overestimate for $\int_{0}^{3} e^{\frac{1}{x}} dx$ with a maximum error less than 0.4.   Give the estimate, the size of n and tell why your estimate is larger than the value of $\int_{0}^{3} e^{\frac{1}{x}} dx$.

III. The purpose of this procedure is to find a method for

predicting the number of subintervals necessary to insure that the maximum error is less than a specified number. To do so, we will find a function whose independent variable, x, is the number of subintervals and the dependent variable y is a bound for the error. Store the values of n and the corresponding values of the bounds for the error from Table 2 in appropriate lists in the stat mode and draw a scatter plot for the error bound vs. the number of subintervals. (see Appendix 1.5) Fit a curve of the form $y = \dfrac{k}{x}$ to the points on the scatter plot by using power regression. (see Appendix 7.8).

1.   List the error function that you fit to the points in the scatter plot and print out a graph (see Appendix 4.1) of this function overlaid on the scatter plot **G1**.

2.   Use this function to estimate the number of subintervals necessary to underestimate $\int_{.5}^{3} e^{1/x} dx$ with an error less than 0.1. Record this value of n.

3.   Store $y = e^{1/x}$ at y1 on the y= menu and execute program RLAREA. Use the value of n in #2 for the number of subintervals. Give the underestimate and a bound for the error. Tell if your estimate is a LHS or a RHS.

4.   Did the function that you found above produce the correct results? Explain.

IV.

1.   Use the program RLAREA with n=15 to estimate $\int_{-2}^{2} \sqrt{4-x^2}\, dx$ with a left-hand sum and a right-hand sum, and give the maximum possible error.

2.   Graph $y = \sqrt{4-x^2}$ and identify this curve as a special geometric shape.

3.   Calculate $\int_{-2}^{2} \sqrt{4-x^2}\, dx$ exactly by using a formula to find the area of the region determined by this definite integral.

4.   Record the error between the estimate in #1 and the exact value in #3.

5.   Does the maximum possible error in #1 agree with the actual error in #4? If not, explain the apparent contradiction.

Checklist of calculator graph printouts to be handed in:

☐   **G1**   Print out a graph of this function overlaid on the scatter plot.

Objectives:
1.   Compute the amount of work done in compressing a cylinder
      by using a program to numerically calculate left- and
      right-hand sums.
2.   Compute the amount of work done in compressing the
      cylinder using the built-in program fnInt on your
      calculator.
3.   Calculate the Contribution to the Mean Effective Pressure
      (MEP_{comp}) from the compression stroke.

Technology:
      TI-82, TI-85 (CBL compatible version), or TI-92 graphing
      calculator, CBL, CBL Pressure Sensor (with syringe), DIN
      adaptor, and programs PRESSURE and RLAREA.

Prerequisites:
1.   Introduction   to   definite   integrals   and   their
      approximations by left- and right-hand sums.

2.   An interpretation of $\int_a^b f(x)\,dx$ as the area of a region.

Overview: Internal combustion engines derive their power from
the expansion of an air-fuel mixture upon combustion. Before
this combustion occurs, the air-fuel mixture is compressed.
This compression process requires work, and the amount of this
work will reduce the power output of the engine commensurately.
A simplified P-V diagram for the four processes of a four-
stroke internal combustion engine is shown in Figure 1.   The
figure actually contains four separate curves: and expansion
curve, an exhaust / intake curve, a compression curve, and a
combustion curve.   We will consider only the expansion function
and the compression function in this project.

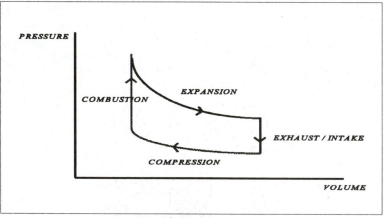

Figure 1

For an ideal engine, the net amount of work done in a complete
cycle is equal to the amount of work done in the expansion
process minus the amount of work required to compress the
cylinder (see Figure 2).

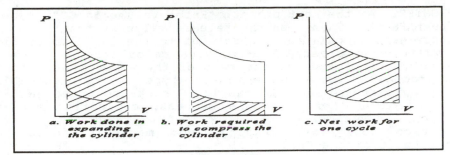

Figure 2

We will simulate the compression stroke of an internal combustion engine using the CBL and a pressure sensor probe (see Figure 3).

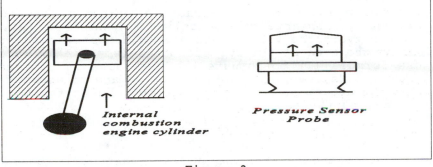

Figure 3

The simulation data will then be used to approximate the amount of work required for compression.

Procedures:
I. Connect the CBL unit to your calculator using a unit-to-unit link cable. Be sure to push both plugs in firmly. Using a CBL DIN adaptor, connect the pressure sensor to the CH1 input port on the top of the CBL unit. Connect the short piece of plastic tubing at the end of the syringe to the three-way valve, with the volume markings facing upward. See Figure 4.

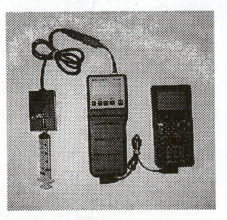

Figure 4

Turn on your calculator and CBL and run the program PRESSURE. When the prompt appears: "How many data points?", type 8 and push ENTER. At the next prompt: "Enter volume in CC", type 20 and push ENTER. Then open the release valve at the top of the pressure sensor, pull the plunger out to the 20 CC mark, close the valve, and press ENTER on your calculator. Leave the release valve closed for the rest of the experiment. At the next prompt: "Enter volume in CC", type 18 and press ENTER. Slowly move the plunger to the 18 CC mark and press ENTER on the calculator. Continue this process, collecting pressure readings for volumes of 16, 14, 12, 10, 8, and 6 CC. After collecting your last reading, you may disconnect your calculator from the CBL.

II. The amount of work done by a force F acting on an object as the object is moved from x = a to x = c is given by:

$$(1) \quad W = \int_a^c F(x)\, dx$$

We could convert the pressure readings to force readings by multiplying by the area of the cylinder head (F = PA). We could also convert the volume readings to distance readings by dividing by the area of the cylinder head (x = V/A). We could then use formula (1) to compute the work done in compressing the cylinder. However, the two conversions are unnecessary for computing the amount of work since the conversion factors will cancel, leaving the equation:

$$(2) \quad W = \int_{V1}^{V2} (PA)\left(\frac{dV}{A}\right) = \int_{V1}^{V2} P\, dV \ ,$$

where V1 corresponds to the volume when x = a and V2 corresponds to the volume when x = c. We need only check that our units match. Units for pressure are Pascals (1 Pa =1 N/m^2). Even though you entered the volume in CC, the PRESSURE program automatically converted these values to cubic meters, and stored them in L_2. With these units, our work will have units of N·m.

1. The PRESSURE program should now be displaying a scatter plot of pressure in Pascals (L_3) vs. Volume in m^3(L_2). Your plot should look similar to the plot in Figure 5. To approximate the amount of work done in N·m in compressing our plunger, we will use left hand sums, as shown in Figure 6. Using the tracer to determine the height of each rectangle, calculate the left-hand sum for the data.

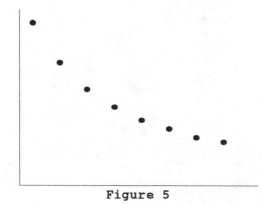

Figure 5

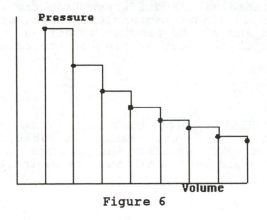

Figure 6

Is this an overestimate or underestimate of the theoretical amount of work?

2. Using right-hand sums, approximate the amount of work done (in N·m) in moving the plunger from the 20 CC mark to the 6 CC mark.

Is this an overestimate or underestimate of the theoretical amount of work?

3. If the temperature of the air were constant throughout compression, and if there were no heat loss in the process, then Boyle's Law would predict that our data should be of the form P=k/V. However, the temperature of the air actually increases significantly during the compression. In fact, the temperature of the air-fuel mixture in your car engine's cylinders may increase to the point where the gas ignites before the spark plug fires. This phenomenon of premature combustion may cause a knocking noise in your engine. Because of the temperature change, a power function of the form $P=ab^V$ fits the data better than the function P=k/V. Use the STAT mode to fit a power function to the Volume (L_2) and pressure (L_3) data, with Volume as the independent variable. See Appendix 7.8 for details on power regression. **G1** Print this function overlaid on a scatter plot of the data (see Appendix 4.1). Record your equation.

4. Now that we have an analytical formula for P = f(V), we are no longer limited to using eight rectangles to approximate $\int_{V1}^{V2} PdV$. Store this formula for P = f(V) in y_1 and run the program RLAREA. The program automatically computes left and right hand sums for a specified number of rectangles. At the prompts, enter $6*10^{-6}$ (this is in m³) $20*10^{-6}$, and 8 (using eight subintervals). Choose option 1 to see a picture of the rectangles for the left-hand sum. Press ENTER to get the value for the left-hand sum. Press ENTER again to see a picture of the right-hand sums. Continue pressing ENTER to work your way through the program. Record the values for the left and right hand sums in Table 1.

5.   Why are these values different from the values that you computed in Part 1 using 8 rectangles? (Look at the height of your rectangles and of the program's rectangles on a graph of P=f(V) overlaid on the scatter plot of the original data.)

6.   Run the RLAREA program again, this time using 16 subintervals and record your results in Table 1. Repeat this once more for 100 subintervals. What happens to the difference between the left- and right-hand sums as we increase the number of subintervals?

7.   Theoretically, what do you think would happen to this difference if we used 1 billion rectangles? Explain your answer.

Number of Subintervals	Left-hand Sum	Right-hand Sum	LHS-RHS
8			
16			
100			

Table 1

8.   Based upon these results, what is your best guess for $\int_{V1}^{V2} P\,dV$?

9.   Your calculator has a built-in program which will approximate $\int_{V1}^{V2} P\,dV$. The format is fnInt(y_1, x, V1, V2). See Appendix 8.4 for details on using fnInt. Using this built-in feature on your calculator, approximate $\int_{V1}^{V2} P\,dV$.

III.   One way to provide a comparative measure of the performance of an engine is to introduce the concept of Mean Effective Pressure (MEP). The MEP is a fictitious pressure that, when multiplied by the displacement volume of the cylinder during the cylinder's power stroke, will produce the same net work as that provided by the actual cycle. That is,

$$MEP \equiv \frac{net\ work\ of\ the\ cycle}{cylinder\ displacement\ volume}$$

Considering only the compression portion of the cycle, we can define the contribution to the MEP from the compression process as

$$MEP_{comp} \equiv \frac{\text{work of compression}}{\text{cylinder displacement volume}}$$

$$= \frac{\int_{V1}^{V2} P\,dV}{\int_{V1}^{V2} dV} = \quad = \frac{\int_{V1}^{V2} P\,dV}{V2-V1}$$

where V1 corresponds to the smallest volume and V2 to the largest volume during the piston stroke.

1.   Approximate the MEP_{comp} for the data that you collected.

2.   Store this value at y_2 and graph y_1 and y_2 on the same viewing screen. **G2** Print this graph. Recall that P=f(V) is stored at y_1, and y_2 is the value of the $M E P_{comp}$. On your printout of the graph, shade the area corresponding to $\int_{V1}^{V2} MEP\,dV$.

3.   How does this area compare to $\int_{V1}^{V2} P\,dV$ which you calculated in procedure II.9?

4.   Why do you think the MEP is called the **mean** effective pressure?

Checklist of graphs and tables to hand in:
☐   **G1**   Power function for cylinder compression (derived from the STAT mode of a graphing calculator.)

☐   **G2**   Graph of MEP_{comp} on same screen as the pressure function.

Objectives:

1.   Investigate the relationship between f(x) and F(x) where
       $F(x) = \int_0^x f(t)\, dt$.

2.   Use a function grapher to graph F(x) where $F(x) = \int_0^x f(t)\, dt$.

3.   Use the graph of F(x) to "guess" a function name for F(x).

Technology:
       TI-82, TI-85, or TI-92 graphing calculator, program CRLIST

Prerequisites:

1.   An introduction to the definite integral $\int_a^b f(x)\, dx$ as the
       limit of a Riemann sum.

2.   An interpretation of $\int_a^b f(x)\, dx$ as the area of a region.

Overview: If $f(t) \geq 0$ for $a \leq t \leq b$, the definite integral $\int_a^b f(t)\, dt$
represents the area bounded by y=f(t), the t-axis, t=a, and t=b.
For x in the interval [a,b], $F(x) = \int_a^x f(t)\, dt$ represents the area of
the shaded region in Figure 1.

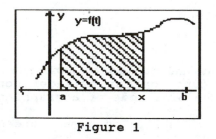

Figure 1

There are several numerical techniques for approximating $\int_a^x f(t)\, dt$
for particular values of x. Your calculator has a built-in
process for estimating F(x) and the command is fnInt. The syntax
of fnInt is fnInt(f(t),t,a,x), where the parameters come directly
from the definite integral notation $\int_a^x f(t)\, dt$, and the order in
which they are written is crucial.

The purpose of this project is to find a relationship between
$F(x) = \int_0^x f(t)\, dt$ and f(x). You will use the fnInt command to
approximate the graph of y=F(x). This graph will be analyzed to
find a formula for F(x). In order to discover the relationship
between F(x) and f(x), you must find a function which "fits" the
data generated by the fnInt command. There are two methods for
fitting a curve to a set of data. One way is to guess a function
and make adjustments until you get a "good fit". This method is

illustrated in Example 1. A second method is to use the curve fitting features of a graphing calculator. This method is illustrated in Example 2.

Example 1: Consider the function f(t)=t. To find a formula for $F(x) = \int_0^x t\, dt$, set up a friendly window (see Appendix 2.5) with xmin=0 and Δx=.05. Graph this function by storing fnInt (T,T,0,x) at y1= ($\int$ (T,T,0,X) on the TI-92). (See Appendix 8.4 for more information about fnInt or $\int$ (integrate).) Your graph should look similar to the one in Figure 2. This appears to be a parabola whose vertex is (0,0). We will use $y=x^2$ as an initial guess for the function name of this graph. The graph of $y=x^2$ overlaid on the graph of y=fnInt(T,T,0,x) is shown in Figure 3.

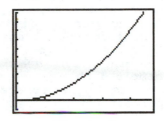

Figure 2

Even though $y=x^2$ doesn't provide a good fit, the function may belong to the family of curves $y=ax^2$ for some value of a. Let's try to find a better value for a. Tracing the curve in Figure 2 shows that the point (1, .5) lies on the graph. Substituting these values for x and y into the equation $y=ax^2$ leads to $0.5 = a(1)^2$, so a = 0.5. Hence, we get the function $y = 0.5x^2$. Check this result by overlaying the graph of $y_2 = 0.5x^2$ on the graph of y_1 = fnInt(T, T, 0, x). The graphs appear to be identical. If you trace to a point on the graph of y_1 = fnInt(T, T, 0, x) and then switch to the graph of y_2 = $.5x^2$, the coordinates should be the same with respect to the accuracy of the calculator.

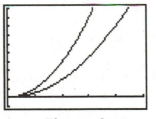

Figure 3

It is rather easy to find a "fit" for the function in this example. It may be more difficult to find an appropriate name for some of the functions in this project. If you experience difficulty with problems 1-5, use the curve fitting technique provided by the statistical features of a graphing calculator. Prior to using these features, the coordinates of the points on the graph must be stored in appropriate lists. To do so, store the function y=fnInt(f(t),T,0,x) at y1 on the y= menu. Run the program CRLIST, which stores x values between 0.01 and 7 in L1 and y values in L2. After successfully executing this program, use power or exponential regression to find an appropriate name for the graph. See Appendices 7.7 and 7.8 for more details about regression. This technique will not work for the three trigonometric functions.

Example 2: Consider the function $F(x) = \int_0^x t \, dt$ whose graph is given in Figure 2. We will use a curve fitting technique available on a TI graphing calculator to find a name for $F(x) = \int_0^x f(t) \, dt$. Set up a friendly window as in Example 1 and store fnInt(T,T,0,x) at y1 on the y= menu. We will now fit a curve of the form $y=ax^b$ to the data generated by $F(x)$. To fit a curve to a set of ordered pairs, the x values must be stored in a list and the corresponding y values in a second list. To save time, the program CRLIST will store all the x values from xmin to xmax in list L1 and the corresponding y-values in list L2. At this time, down load program CRLIST from your program disk and run this program with FnInt(T,T,0,x) stored at y1. Now use Appendix 7.8 as a guide to fit a curve of the form $y=ax^b$ (power regression) to this data. When you complete this process you should see $y=ax^b$ where a=.5, b=2, r=1. This is interpreted as $y=.5x^2$. The r=1 means that this is a perfect fit.

Procedures:

I.   For each of the functions $F(x)$, where $F(x) = \int_0^x f(t) \, dt$, use Example 1 or Example 2 as a guide to find a function name for $F(x)$. Use the same window as in Example 1 to first graph $F(x)$. Fill in Table 1 with your results.

1. $F(x) = \int_0^x 2 \, dt$ 2. $F(x) = \int_0^x t \, dt$

3. $F(x) = \int_0^x t^2 \, dt$ 4. $F(x) = \int_0^x t^3 \, dt$

5. $F(x) = \int_0^x t^4 \, dt$ 6. $F(x) = \int_0^x \cos t \, dt$

7. $F(x) = \int_0^x \sin t \, dt$ 8. $F(x) = \int_0^x \sin 2t \, dt$

9. $F(x) = \int_0^x e^t \, dt *$ 10. $F(x) = \int_0^x e^{2t} \, dt$

* (compare this graph with $y=e^x$)

$F(x) = \int_0^x f(t)\ dt$	
f(t)	Function name for F(x)
1. f(t) = 2	
2. f(t) = t	
3. f(t) = t^2	
4. f(t) = t^3	
5. f(t) = t^4	
6. f(t) = cos t	
7. f(t) = sin t	
8. f(t) = sin 2t	
9. f(t) = e^t	
10. f(t) = e^{2t}	

Table 1

II.   Analyze the results in Table 1 to find a calculus
relationship between F(x) and f(x) involving the derivative.
State the relationship as a theorem.   Consult your calculus book
and find the name of this theorem.

Objectives:
1.   Observe a situation where the analytical model of motion
      due to gravity is a poor model.
2.   Derive a better model for motion which accounts for the
      effect of air resistance.
3.   Verify the accuracy of the new model by comparing it with
      experimental data from the CBL.

Technology:
      TI-82, TI-85 (CBL compatible version), or TI-92 graphing
      calculator, CBL, CBL motion detector, programs BALLDRPX,
      CHOOSE, and VELOCITY

Equipment:
1.   Beach ball
2.   Two rubber bands
3.   Rod and clamp (see Figure 2)

Prerequisites:
      Introduction to rectilinear motion and separable
      differential equations.

Overview: The analytical model for the motion of a falling body,
$y=at^2+bt+c$, neglects the effects of air resistance. In many
situations this model is reasonably accurate. However, for an
object with a large cross-sectional area relative to its weight,
air resistance can affect the motion significantly. We will
investigate this behavior by analyzing the motion of a beach
ball.

Procedures:
I.   A beach ball is dropped from the balcony of a tall
building. Assign a reference system as shown in Figure 1, with
y = 0 the initial position of the ball, and y directed downward.
Find an equation for the position of the ball (in feet) as a
function of time, y = y(t), and the velocity of the ball (in
feet/sec) as a function of time, v = v(t), in two different ways:

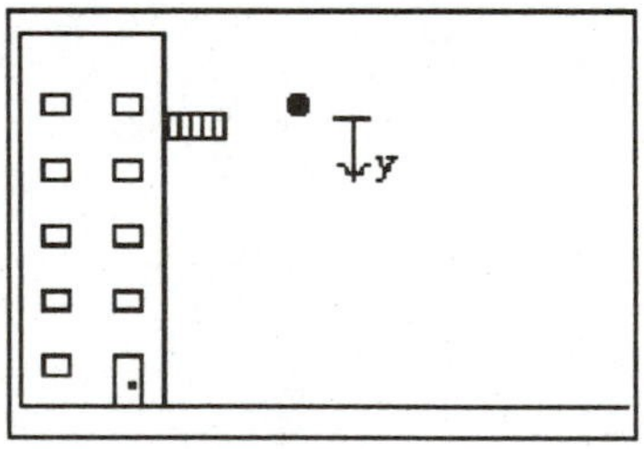

Figure 1

A.   Assume that air resistance is negligible. Record your
formulas for position and velocity.

B.   Assume that the force of air resistance is equal to kv,
where k is a constant and v is the velocity of the ball. You
will need to know the weight of your beach ball. If you don't
have a scale, use a weight of 0.2 pounds. Using Newton's Second
Law,

$$m\frac{dv}{dt} = 32m - kv$$

Here m is the mass of the beach ball (weight = 32m), and k is the constant corresponding to the air resistance. Separating variables, we have

$$\frac{mdv}{32m - kv} = dt$$

Keeping in mind that k is constant, antidifferentiate to obtain velocity as a function of time, v = v(t). Use the initial velocity (at t=0) to solve for the constant of integration. Antidifferentiate a second time to get the position of the ball as a function of time, y = y(t). Use the initial position (at t=0) to solve for the constant of integration. Record your equations for position and velocity.

II. To test the effectiveness of the two models, experimental data will be collected with a motion detector and CBL. To observe the effects of air resistance on the motion of the ball, it must be dropped from a height of at least 10 feet, but not more than 20 feet, since the motion detector cannot detect distances larger than about 19-20 feet. The balcony should have a secure railing, and the path of the ball should be free of drafts and air currents.

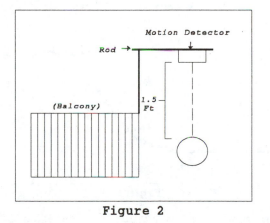

Figure 2

Attach the motion detector to a rod with two rubber bands, and then attach the rod to the railing so that it extends out over the balcony as shown in Figure 2. Extend the rod so that the motion detector is about 2 feet beyond the balcony, or just far enough to safely reach from the balcony. Turn on the CBL. With the calculator attached to the CBL by means of a unit-to-unit link cable, and the motion detector plugged into the sonic port of the CBL, start the BALLDRPX program. See Figure 3 for CBL hookup.

Figure 3

Press the trigger button on the CBL and within a second, release
the beach ball. The BALLDRPX program records the position of the
ball every three hundredths of a second. Do not attempt to drop
the ball at exactly the same time that the trigger button is
pressed; because readings are being taken every three hundredths
of a second, human error will produce inconsistent results. If
your scatter plot doesn't look similar to Figure 4 in the region
where the ball is falling (between T_0 and T_F), repeat the data
collection process.

III. To analyze the results, first run the CHOOSE program. This
program automatically chooses only the portion of the curve when
the ball is actually falling (see Figure 4). The CHOOSE program
will "throw out" all points except those between the points t_0
(where the ball is released) and t_f (where the ball hits the
ground). Time values are stored in L_1 and distance values are in
L_2.

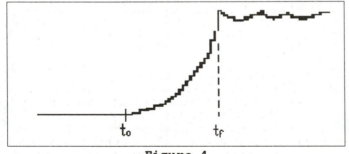

Figure 4

The CHOOSE program will also adjust the data to fit our reference
system. Recall that we want the ball to start at $y = 0$, but we
must start with the ball approximately 1.5 feet below the motion
detector (it cannot detect objects closer than that). So the
CHOOSE program "shifts" the data to match our reference system
with an initial distance of zero.

IV.   A. We will now try to fit this adjusted experimental data
with an analytical model. First, overlay the graph of the
function $y = y(t)$ which you obtained from IA on a scatter plot of
the experimental data points (see Appendix 1.6). This is the
model we obtained by neglecting air resistance. Observe that
this model does a very poor job of describing the motion of the
beach ball. **G1** Store this picture to be printed later. (See
Appendix 10.1).

B. Next we will see how well our analytical model $y = y(t)$ from
IB fits the data. This is the model which assumes an air
resistance force of kv. In order to graph this function, we need

to assign a value for k. The value for k depends on many factors, including the size, weight, and shape of the ball, as well as the viscosity of the air. We will choose values of k to fit the data, starting with an initial guess, and changing the value of k until the model fits the data.

1.   Enter the formula for the function $y = y(t)$ at the y= menu by typing in "k" wherever it appears in the formula. That is, do not type in a value for k at this point.

2.   Exit the "y=" mode and store a value for k such as k=.001 (Press .001 **[STO]** k).

3.   Overlay the graph of the function from (1) on a scatter plot of the experimental data.

4.   Based on how well the model fits the experimental data, choose a different value for k and repeat steps (2) and (3) until you have a "good fit." Record your value for k.

Observe how well this model fits the data. **G2** Store this picture to be printed later.

The value for k which you found depended on the properties of the beach ball and the air. This value for k will remain valid for other experiments involving motion of the same ball through the same air, provided that the velocities are in approximately the same range.

C.   Even though $y=at^2+bt+c$ is not a reasonable model for the motion of a beach ball for y<20 ft, it is a reasonable model for the motion of a basketball for y<20 feet. Discuss why the model is reasonable in one case but not the other.

V.   To obtain information about the velocity of the ball, run the VELOCITY program. The program computes the average velocities over each time interval of 0.03 seconds and stores the velocities in L_5. The program will automatically draw a scatter plot of these experimental velocity values (L_5) vs. Time (L_1). On the same screen plot the graph of the analytical model $v = v(t)$ from IA by entering your formula for v at the y= menu. This is the model which assumes that air resistance is negligible. Observe that the model is not very accurate. **G3** Store the picture to be printed later.

Next plot the analytical model of $v = v(t)$ from IB on the same screen as the experimental velocities. This model assumed that the air resistance force was kv. Use the value for k that you found in IVB. **G4** Store this picture to be printed later.

1.   Does the analytical model accurately predict the velocity of the ball?

2.   Give possible explanations for any differences between the graph of this analytical model, which accounts for air resistance, and the experimental results.

VI.   Printout the graphs **G1**-**G4** below (see Appendix 10.2 to recall each picture and Appendix 4.1 to print them).

Checklist of calculator graph printouts to be handed in:

☐   **G1**   Experimental data and analytical model (neglecting air resistance) of the position of the ball as a function of time - Procedure IV A.

☐   **G2**   Experimental data and analytical model (accounting for air resistance) of the position of the ball as a function of time - Procedure IV B.4.

☐   **G3**   Experimental data and analytical model (neglecting air resistance) of the velocity of the ball as a function of time - Procedure V.

☐   **G4**   Experimental data and analytical model (accounting for air resistance) of the velocity of the ball as a function of time - Procedure V.

Objectives:
1.   Introduce economic concepts of consumer surplus and producer surplus, and the approximation of these quantities by integration.
2.   Work through a problem requiring a variety of mathematical techniques and some creativity.

Technology:
     TI-82, TI-85 or TI-92 graphing calculator

Prerequisites:
1.   Introduction to integration as a method of calculating area.
2.   Evaluation of definite integrals of polynomial functions.

Overview:
In economics the equilibrium quantity q* is defined to be the quantity at which the supply equals the demand. See Figure 1.

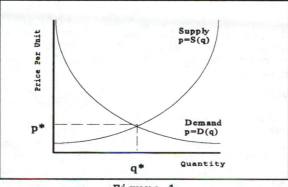

Figure 1

The corresponding price p* is referred to as the equilibrium price. The consumer's surplus is then defined to be the total savings to consumers who are willing to pay a higher price than the equilibrium price. The consumer's surplus is the area shaded in Figure 2.

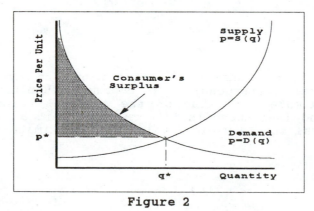

Figure 2

This is a savings to consumers since they will pay the equilibrium price, p*, rather than the higher price (on the Demand curve) which they were willing to pay.

The producers' surplus is defined to be the total gain to producers who are willing to supply the product at a price lower than the equilibrium price. This is the area shaded in Figure 3.

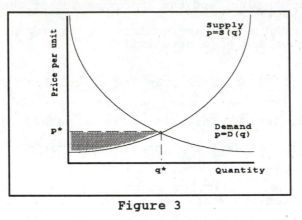

Figure 3

Procedures:

Nancy has compiled all of her notes from a meeting with the marketing director of Wacky Wickets Corporation and realizes that she is missing some information. She has the following facts written in her notes:

Supply equation is $p=S(q)=10+\dfrac{1}{10}q+\dfrac{1}{3600}q^2$

Consumer surplus is $200,000
Producer surplus is $132,000

Unfortunately, it is the demand equation which she is most interested in, and she does not have it written in her notes. She recalls that the marketing director used a linear function to model the demand function, but does not have the equation written down.

1. Find the linear function $p=D(q)=aq+b$ which was used to obtain the information given above. You may use any technique to obtain an estimate for the parameters a and b.

2. Confirm your answer by setting up and evaluating the integral corresponding to consumer surplus. You may approximate the value of the integral using the fnInt command (see Appendix 8.4) or compute it directly. This integral should yield a value of approximately $200,000.

Objectives:
1. Approximate a definite integral using a sequence of trapezoids.
2. Explore the relationship between the value of a definite integral and the sum of the areas of trapezoids.
3. Devise a method for finding a bound for the error when the sum of the areas of a sequence of trapezoids (secant type) or the sum of the areas of a sequence of trapezoids (tangent line type) are used to approximate a definite integral.
4. Develop a method for estimating the number of subintervals necessary to approximate a definite integral with a sum of the areas of trapezoids for a specified error bound.

Technology:
TI-82, TI-85 or TI-92 graphing calculator, program MTAREA.

Prerequisites:
1. Introduction to the definite integral as a limit of Riemann Sums.
2. Introduction to concavity of a function and points of inflection.
3. Introduction to the Fundamental Theorem of Calculus.

Overview: The Fundamental Theorem of Calculus is a powerful method of evaluating definite integrals such as $\int_{0}^{\pi/2} \sin x \, dx$.

The power of this theorem to evaluate a definite integral, $\int_{a}^{b} f(x) \, dx$, depends on one's ability to find an antiderivative of $f(x)$. Definite integrals such as $\int_{0}^{2} \sin\sqrt{x} \, dx$ or $\int_{.5}^{3} e^{1/x} dx$ elude the Fundamental Theorem of Calculus because it is difficult to find an antiderivative for either integrand. There are many techniques available for approximating such integrals. When you use an approximation, you are usually expected to tell how "good" the approximation is by finding a bound for the error.

The purposes of this project are:
1. Define an appropriate sequence of trapezoids so that the sum of the areas of these trapezoids approximates the value of a definite integral.

2. Develop a procedure to find a bound for the error when using the sum of the areas of a sequence of trapezoids to approximate the value of a definite integral.

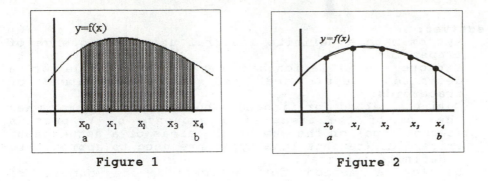

Figure 1   Figure 2

Consider an integral $\int_a^b f(x)\,dx$ and the associated region which is displayed in Figure 1.

The interval from a to b is subdivided into four subintervals each of width $\Delta x = \dfrac{b-a}{4}$. The region associated with the integral $\int_a^b f(x)\,dx$ and the four trapezoids are displayed in Figure 2.

The area of a trapezoid is $\left(\dfrac{B+b}{2}\right)h$, where B and b represent the lengths of the parallel bases and h is the height. The area for the first trapezoid is $\left(\dfrac{f(x_1)+f(x_0)}{2}\right)\Delta x$, where the lengths of the bases are $f(x_1)$ and $f(x_0)$ and the "height" of this trapezoid is $\Delta x = x_1 - x_0$. The sum of the areas of the four trapezoids is $\displaystyle\sum_{k=1}^{4}\left(\dfrac{f(x_k)-f(x_{k-1})}{2}\right)\Delta x$ and is called TS(4).

From Figure 2, TS(4) appears to be an underestimate for $\int_a^b f(x)\,dx$. The concavity of a function will determine whether these sums are underestimates or overestimates. If a function f is concave down over [a,b] then TS(4) is an underestimate for $\int_a^b f(x)\,dx$ (See Figure 3). If the function is concave up over [a,b] then TS(4) is an overestimate for $\int_a^b f(x)\,dx$ as illustrated in Figure 4.

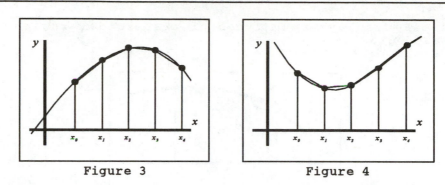

Figure 3   Figure 4

In order to find a bound for the approximating error, we

need a way to calculate an overestimate for $\int_a^b f(x)\,dx$ when

TS(n) is an underestimate and an underestimate when TS(n)
is an overestimate. If f(x) is concave down on an
interval [a,b] containing c, then the curve lies below
the tangent line at c. See Figure 5. We can use this
tangent line to form the trapezoid which is shown in
Figure 5.

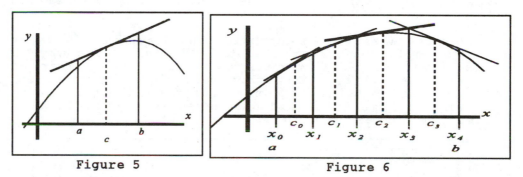

Figure 5   Figure 6

In Figure 6, trapezoids are formed by constructing
tangent lines at c_0, c_1, c_2, and c_3 which are midpoints of
the four subintervals. The sum of the areas of these

trapezoids is an overestimate of $\int_a^b f(x)\,dx$ for the function

in Figure 6. Look at Figures 7 and 8 and notice the
difference between the trapezoids depicted in Figures 7
and 8.

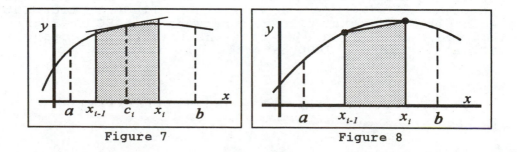

Figure 7   Figure 8

Figure 9 displays both types of trapezoidal regions in the same graph.

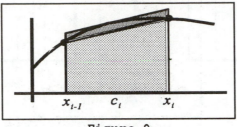

Figure 9

If a function is concave down over [a,b] as shown in Figures 7 and 8, a sum of areas of tangent line trapezoids is an overestimate and a sum of areas of secant line trapezoids is an underestimate. If a function is concave up over [a,b] such as the one in Figures 10 and 11, the sum of the areas of the tangent line trapezoids is an underestimate and the sum of the areas of the secant line trapezoids is an overestimate.

One must be careful when approximating a quantity because of round off error from the calculations. One way to reduce the overall error is to reduce the number of computations involved in determining the approximation. More calculations are necessary for finding the area of the tangent line trapezoid displayed in Figure 10 than the secant trapezoid displayed in Figure 11.

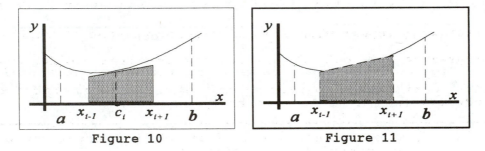

Figure 10   **Figure 11**

To reduce the number of calculations for computing the area of a tangent line trapezoid we will find the area of a rectangle that has the same area as the trapezoid. Choose a rectangle whose height is $f(c_i)$ where c_i is the midpoint of $[x_i, x_{i+1}]$. This is called a midpoint rectangle. Note that $\triangle ABC$ is congruent to $\triangle DEC$ in Figure 12. Since these triangles have the same area, rectangle PBEQ and trapezoid PADQ have the same area.

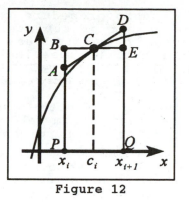

Figure 12

Hence we can use the sum of the areas of midpoint rectangles to calculate the sum of the areas of tangent line trapezoids.

Procedures:
I.
1. Store $f(x) = e^{\frac{1}{x}}$ at Y_1 on the y= menu and graph this function over the interval [.5,3]. Based on this graph, explain how to find an underestimate and an overestimate for $\int_{.5}^{3} e^{\frac{1}{x}} dx$.

2. The program, MTAREA, calculates the sum of the areas of secant line trapezoids and the sum of the areas of midpoint rectangles associated with an integral such as $\int_{.5}^{3} e^{\frac{1}{x}} dx$. The function $f(x) = e^{\frac{1}{x}}$ <u>must</u> be stored at Y_1 or an error will appear when the program is executed. Use this program to fill out Table 1 for the integral $\int_{.5}^{3} e^{\frac{1}{x}} dx$. Use the difference between an overestimate and an underestimate to calculate an error bound.

n	sum of areas of midpoint rectangles	sum of areas of secant line trapezoids	error bound
5			
10			
20			
40			
60			

Table 1

3. Describe the behavior of the error bound as n increases.

II. The purpose of this procedure is to find a method for predicting the number of subintervals necessary to insure than an error bound is less than a specified number. We want a function whose independent variable, x, is the number of subintervals and the dependent variable y is a bound for the error.

1. Store the values of n and the corresponding values of the bounds for the error from Table 1 in lists L1 and L2 and draw a scatterplot for error bound vs. the number of subdivisions. (See Appendix 1.5). Use a trial and error approach to fit a curve of the form $y = \dfrac{k}{x^2}$ to the points on the scatterplot. Try a value of k and overlay this graph on the scatterplot. Make adjustments until your "eyeball" fit looks <u>real</u> good. **G1** Print out a picture (see Appendix 4.1) of the graph of this function overlaid on the scatter plot (see Appendix 1.6), and record this function.

2. Use this function to estimate the number of subintervals required to approximate $\int_{.5}^{3} e^{\frac{1}{x}} dx$ with an error less than .001.

3. Find an underestimate and an overestimate for $\int_{.5}^{3} e^{\frac{1}{x}} dx$ using the value of n in question 2. (Use program MTAREA with $y1 = e^{1/x}$).

4. Use the results in #3 to explain whether or not the n in #2 is appropriate.

III. Up to this point in this project we have considered integrals $\int_{a}^{b} f(x)\, dx$ such that f was either concave up or concave down over the entire interval [a,b]. In such cases, the actual value of the integral, A, is between the sum of the areas of tangent line trapezoids and the sum of the areas of secant line trapezoids for any n. The purpose of this procedure is to develop a method for finding an overestimate and an underestimate for $\int_{0}^{1.75} \sin(x^2)\, dx$ and then calculate an error bound. The techniques of Procedure I do not apply directly to the integral $\int_{0}^{1.75} \sin(x^2)\, dx$ because $f(x) = \sin(x^2)$ is not strictly concave up or strictly concave down over [0,1.75]. However, the techniques of Procedure I can be applied on appropriate subintervals of [0,1.75].

1. Explain in detail how to use tangent line trapezoids

and secant line trapezoids to find an overestimate and an underestimate for $\int_0^{1.75} \sin(x^2)\,dx$. Any estimate of x-values which are used in your set-up must be accurate to 3 decimal places.

2.   Using MTAREA, find an estimate for $\int_0^{1.75} \sin(x^2)\,dx$ such that the bound for the error is less than 0.005. Show all work and record any calculations that are used, as well as the value or values of n which you used in the program MTAREA.

Checklist of calculator graph printouts to be handed in:
☐   **G1**   Print out a picture of the graph of this function overlaid on the scatter plot, and record this function.

Objectives:
1.   Develop an understanding of the graphical interpretation of
      the definite integral and the development of the integral
      as the limit of a sum.
2.   Apply the concept of the definite integral to biology and
      physics.
3.   Interpret the mean value theorem for integrals in a
      geometrical application.

Technology:
      None required.

Prerequisites:
1.   Basic introduction to the definite integral.
2.   Working knowledge of summation notation.
3.   Evaluate basic definite integrals by using the fundamental
      theorem of calculus.

Overview: The definite integral is defined as the limit of a Riemann sum, "adding up an infinite number of pieces." For example, we could "add up the areas of an infinite number of rectangles" to obtain the area under the curve y=f(x) from x=a to x=b (see Figure 1). Since it is really not possible to add an infinite number of pieces, we add a finite number of pieces, say n pieces, and take the limit as $n\rightarrow+\infty$. If the function f is continuous on the interval [a,b], then we can write this limit as an integral:

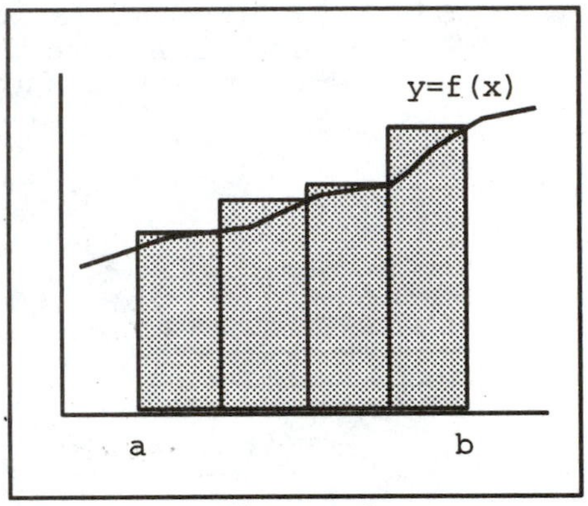

Figure 1

$$A = \lim_{n\rightarrow\infty} \sum_{i=1}^{n} f(x_i)\,\Delta x = \int_a^b f(x)\,dx$$

There are many different ways of representing this area as a limit of Riemann sums but all have the same limit. Here are a few representations:

$$A = \lim_{n\rightarrow\infty} \sum_{i=1}^{n} f(x_i)\,\Delta x$$

$$= \lim_{n\rightarrow\infty} \sum_{i=1}^{n} f(x_{i-1})\,\Delta x$$

$$= \lim_{n\rightarrow\infty} \sum_{i=0}^{n-1} f(x_i)\,\Delta x$$

It is easy to become confused with the subscripts and notation. Our purpose here is to look at the integral in an intuitive sense, so we will be loose with the notation in order to gain a better understanding of the meaning of the integral.

Consider again the area example. For any x value in a particular subinterval, the height of the rectangle corresponding to this subinterval is approximately f(x) and its width is Δx, which can be thought of as dx. So a loose but insightful way of representing the area is:

$$A = \lim_{\Delta x \to 0} \sum f(x)\, \Delta x = \int_a^b f(x)\, dx$$

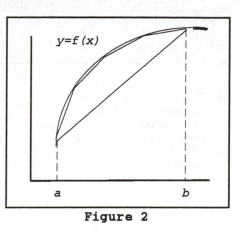

Figure 2

This not only makes it easier to see the formation of the integrand, but also emphasizes the definition of the integral as "adding up all of the pieces." We could also "add up the lengths of an infinite number of line segments" to obtain the length of the arc of the curve y=f(x) from x=a to x=b (see Figure 2).

The length of each of these segments is $\Delta \ell = \sqrt{(\Delta x)^2 + (\Delta y)^2}$ (see Figure 3).

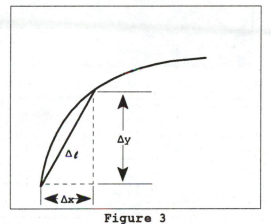

Figure 3

$$\text{arc length} \quad = \lim_{\Delta x \to 0} \sum \sqrt{(\Delta x)^2 + (\Delta y)^2}$$

$$= \lim_{\Delta x \to 0} \sum \sqrt{1 + \left(\frac{\Delta y}{\Delta x}\right)^2}\; \Delta x$$

$$= \int_a^b \sqrt{1 + (dy/dx)^2}\; dx$$

In thinking of the definite integral in this way (i.e. as "adding up an infinite number of pieces") the definite integral can be applied to a virtually endless number of situations. We will explore only a few to establish a pattern, and your imagination and creativity will provide access to others.

Procedures:
I.   Suppose that the velocity of a car is constant at 60 mph for a long trip. To find the total distance traveled in 3 hours we simply use the formula d=vt, and calculate

$$d = \left(60\, \frac{miles}{hour}\right)(3 \text{ hours}) = 180 \text{ miles.}$$

However, if the velocity is not constant, say v=f(t), we cannot apply this formula to easily compute the total distance traveled. But we can use the formula d=vt in conjunction with integration to find the total distance traveled from t=0 to t=3 hours.

1.   Write the total distance traveled as the limit of a sum. See Figure 4. Subdivide the interval from t=0 to t=3 into n subintervals of equal width, Δt. When n is large, the velocity on any subinterval is almost constant. That is, the difference between vmax and vmin on each subinterval is small enough so that the formula d=vt provides a good approximation for the distance traveled over time period Δt.

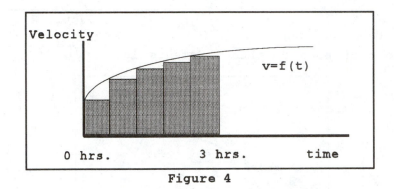

Figure 4

2.   Use this formula to write the total distance traveled from t=0 to t=3 hours as an integral.

II.   Suppose that the acceleration of a car is constant at 2 mph/hour for a long trip and that the car starts with a velocity of 60 mph.   Then the velocity of the car after 3 hours has increased by

$$v = a \cdot t = \left(2\,\frac{mph}{hours}\right)(3 \text{ hours}) = 6 \text{ mph.}$$

Since the initial velocity was 60 mph, the velocity after 3 hours will be 66 mph. In most cases, however, the acceleration is not constant. Suppose that acceleration is given by a=g(t).

3.   By breaking the interval from t=0 to t=3 into n equal subintervals and assuming that the acceleration is almost constant on each of these subintervals, write the total increase in velocity as the limit of a sum.

4.   Use this formula to write the total increase in velocity from t=0 to t=3 hours as an integral.

III. Suppose that the number of bacteria cells in a population is constant at N=1,000,000 cells and that the average consumption rate per bacteria cell (i.e. how much food they eat per hour) is constant at $r = \dfrac{5\,nm^3\ of\ food/cell}{hour}$.   Then the total amount of food consumed by the whole population in 4 hours is

$$\left(\dfrac{5\,nm^3/cell}{hour}\right)(1,000,000\ cells)(4\ hours) = 20,000,000\ nm^3\ of\ food.$$

The consumption rate is usually not constant, though, and we will consider two ways in which the rate may vary.

Case 1: The consumption rate depends on time. This may occur due to large temperature drops which may cause the metabolism and feeding rate to slow. Let $r=f(t)$ represent the consumption rate and $N=g(t)$ represent the number of bacteria in the population at any time t.

5.   Break the interval [a,b] on the time axis into equal subintervals (see Figure 5) and, assuming that the population and consumption rates are approximately constant on each subinterval, write the total food consumption as the limit of a sum.

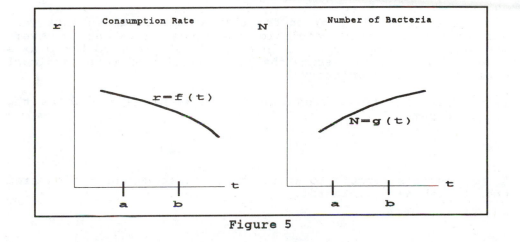

Figure 5

6.   Use this formula to represent the total consumption from t=a to t=b as an integral.

Case 2: The consumption rate depends on the population size. This may occur when the population gets too large for its environment and access to food becomes limited. Also, build up of waste products affect the health, and hence the consumption rate, of the cells. Let $N=g(t)$ represent the number of bacteria cells in the population and let $r=f(N)$ represent the consumption rate. See Figure 6.

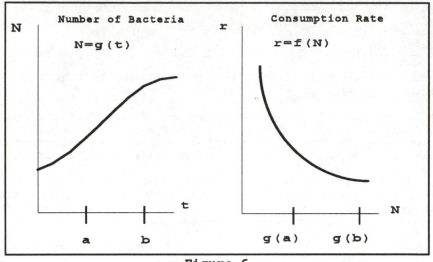

Figure 6

Break the interval [a,b] on the t-axis into equal subintervals and assume that the population is almost constant on every subinterval. Also break the interval [g(a),g(b)] on the N-axis into subintervals and assume that the consumption rate is almost constant on each subinterval.

7. Write the total food consumption from t=a to t=b as the limit of sums.

8. Use this formula to write the total food consumption from t=a to t=b as an integral.

IV. Consider the problem of finding the average depth of the swimming pool in Figure 7, which has three different depths as shown.

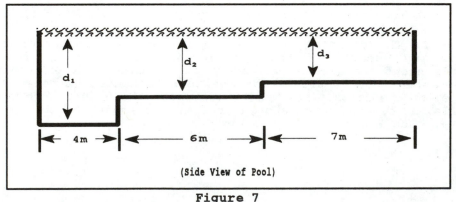

(Side View of Pool)

Figure 7

The average depth cannot be calculated by a simple average of the three depths, $\dfrac{d_1+d_2+d_3}{3}$, since the depths apply to different sized portions of the pool. Instead, we need to find the "weighted average" of the depth as $\dfrac{4d_1+6d_2+7d_3}{4+6+7} = \dfrac{4d_1+6d_2+7d_3}{17}$. This "weighted average" accounts for the fact that the pool has a depth of d_1 for $\dfrac{4}{17}$ of the total length of the pool, d_2 for $\dfrac{6}{17}$, and d_3 for $\dfrac{7}{17}$ of the total length.

However, most pools have an evenly contoured bottom (or else stubbed toes would plague the swimming world). Suppose that the depth of a pool is given by a function d=f(x) meters where x meters is the distance from the shallow end of the pool. See Figure 8.

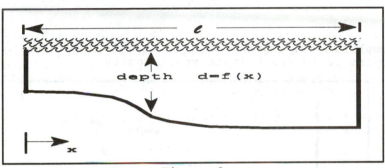

Figure 8

9.   By breaking the length of the pool into small pieces and assuming that the depth is almost constant for each piece, write the average depth as the limit of a sum.

10.   Use this formula to write the pool's average depth as an integral.

11.   Generalize your results about average pool depth to obtain a formula for the average value of a continuous function on [a,b]. See Figure 9.

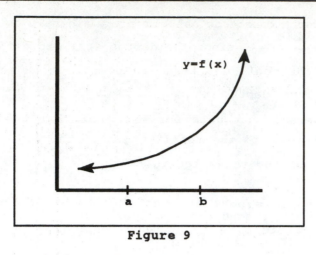

Figure 9

12. Use your formula for the average value to find an expression for the average slope of the continuous function f on [a,b].

13. Simplify your results and interpret them geometrically, using Figure 10 to demonstrate your results.

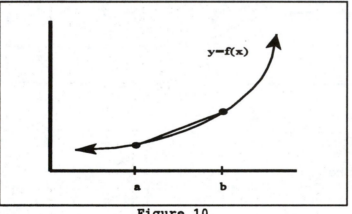

Figure 10

14. The mean value theorem for integrals states that for a function f which is continuous on [a,b] there exists a number α in [a,b] such that

$$\int_a^b f(x)\, dx = f(\alpha)(b-a)$$

Rewrite this equation for our swimming pool example in Figure 8 by replacing a and b with the appropriate expressions. What does $f(\alpha)$ represent in our problem?

15.   What does α represent in our problem?

V.   The <u>centroid</u> of an object is a point which defines the geometrical center of the object. For some flat objects, such as the circular and rectangular objects in Figure 11, locating the centroid is easy due to the symmetry of the object.

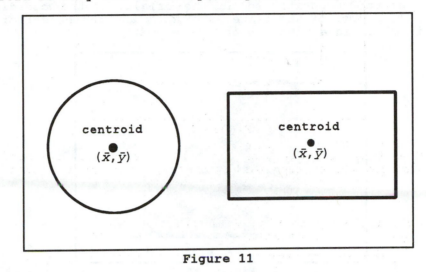

centroid
$(\bar{x}, \bar{y})$

centroid
$(\bar{x}, \bar{y})$

Figure 11

But how does one locate the centroid $(\bar{x}, \bar{y})$ of an object such as the one in Figure 12? We can break the object into three pieces, each of which has a known centroid. We will call these centroids $(\tilde{x}_1, \tilde{y}_1)$, $(\tilde{x}_2, \tilde{y}_2)$, $(\tilde{x}_3, \tilde{y}_3)$ for the first, second and third rectangles, respectively. See Figure 13. Each region also has a known area, A_1, A_2, and A_3, respectively.

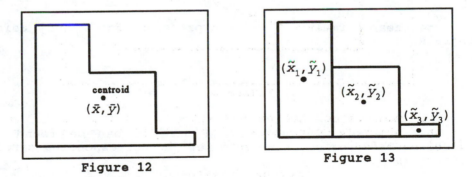

centroid
$(\bar{x}, \bar{y})$

Figure 12

$(\tilde{x}_1, \tilde{y}_1)$

$(\tilde{x}_2, \tilde{y}_2)$

$(\tilde{x}_3, \tilde{y}_3)$

Figure 13

We can now calculate the "weighted average" to find the overall centroid, $(\bar{x}, \bar{y})$, for the composite of the three pieces:

$$\bar{x} = \frac{A_1\tilde{x}_1 + A_2\tilde{x}_2 + A_3\tilde{x}_3}{A_1 + A_2 + A_3}$$

$$\bar{y} = \frac{A_1\tilde{y}_1 + A_2\tilde{y}_2 + A_3\tilde{y}_3}{A_1 + A_2 + A_3}$$

These coordinates, $(\bar{x}, \bar{y})$, locate the centroid of the object shown in Figure 12. If this thin, flat object is made entirely of the same material and has uniform thickness, then the centroid is also the "balancing point" for this object.

Now consider the thin, flat plate shown in Figure 14. The corner of the plate is positioned at the origin of the coordinate system. The width of the plate is b units and the height of the plate is given as a function of x, y=f(x).

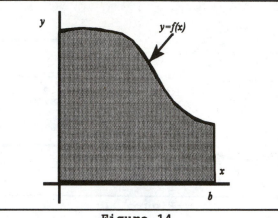

Figure 14

16. By breaking the x-axis into small pieces and assuming that the height of the plate is almost constant on each interval, write the x and y coordinates of the centroid ($\bar{x}$ and $\bar{y}$) as limits of sums.

17. Use these formulas to write expressions for $\bar{x}$ and $\bar{y}$ using integrals.

VI. The area of the region in Figure 15 can be found by considering this region as a sector of the corresponding circle in Figure 16.

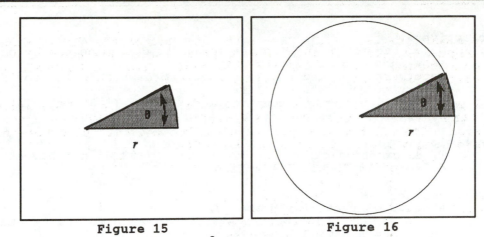

Figure 15   Figure 16

The area of the sector is $\dfrac{\theta}{2\pi}$ of the total area of the circle, where θ is measured in radians.

$$A = \frac{\theta}{2\pi}(\pi r^2) = \frac{1}{2}\theta r^2$$

Consider the region in Figure 17 which is bounded by the graph of a polar function $r=f(\theta)$ and rays $\theta=\alpha$ and $\theta=\beta$, where θ is measured in radians.

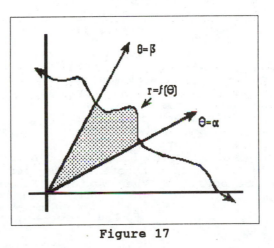

Figure 17

18.   By breaking the angle between α and β into small pieces $\Delta\theta$ and assuming that the radius is almost constant for each of these pieces, write the area of the region as the limit of a sum.

19.   Use this formula to write the area as an integral.

Objectives:
1.   Model the motion of a bungee jumper to determine his height
       as a function of time.
2.   Determine the length of a bungee chord which will provide
       a particular distance of fall.
3.   Use the differential equation mode of the TI-85 calculator
       to solve a non-separable differential equation.

Technology:
       TI-85 graphing calculator (TI-82 and TI-92 do not have the
       differential equations mode.)

Prerequisites:
1.   A study of rectilinear motion.
2.   An introduction to Newton's Second Law of Motion, $\Sigma F=ma$,
       and Hooke's Law, $F=ks$.
3.   An introduction to separable differential equations and
       boundary value problems.

Overview:
Free-Body Diagrams: In order to apply Newton's Second Law of Motion, $\Sigma F=ma$, to an object, we must know what forces are acting on the object. This force information can be summarized on a free-body diagram, in which the object is treated as a rigid dot, and all forces acting on the object are represented by vectors. For example, suppose a 170 pound man is standing in an elevator. As the elevator begins to rise, we can summarize the forces acting on the man with a free-body diagram as

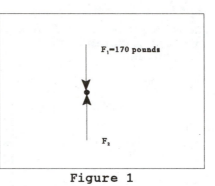

Figure 1

shown in Figure 1. Force F_1 is the force of gravity acting on the man's body and F_2 is the force that the floor of the elevator exerts on his feet. If our reference system defines the upward direction to be positive, then Newton's Second Law is

$$\Sigma F = ma$$
$$F_2 - 170 = \frac{170}{32.2}a$$
$$a = \frac{32.2}{170}(F_2 - 170)$$

where a is the acceleration of the man's body, which is a function of time. In fact, $a = \frac{dv}{dt}$, so

$$\frac{dv}{dt} = \frac{32.2}{170}(F_2 - 170)$$

The purpose of the free-body diagram in this example was to help us visualize the forces acting on the man's body, $\Sigma F = F_2 - 170$.

Bungee Jumping: The "sport" of bungee jumping has dramatically increased in popularity since its introduction to the United States in the 1980's. We will model the motion of a 180 pound bungee jumper by assuming that the bungee chord behaves as a spring with a spring constant of 10 pounds per foot. The jumper begins on a platform which is 200 feet above the surface of a swimming pool. A bungee chord with a free length of 30 feet is

attached to the jumper's ankles. The chord is then connected to
a solid rope 20 feet long, which is tied to the platform. See
Figure 2.

Assume that the air resistance force during the free fall (before
the bungee chord engages) is 0.7 times the magnitude of the
velocity of the jumper, 0.7v. Assume that the damping force
while the chord is engaged is 1.5v. This damping force accounts
for air resistance and for friction in the bungee chord, and
acts in the direction opposing the motion of the jumper as he
falls.

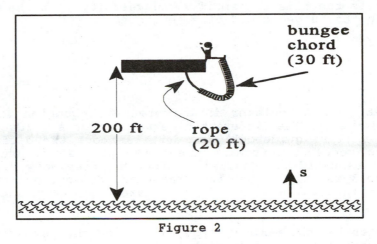

Figure 2

Assume that the mass of the rope and bungee chord is negligible.
Assign your reference system as shown in Figure 1 with the height
s=0 corresponding to the surface of the pool, and the positive
direction upward. (Please note: the formula for motion
$P_y = -\frac{1}{2}at^2 + v_{y_0}t + P_{x_0}$ is not valid here since it assumes that there
are no external forces in addition to gravity.)

Procedures:
1. Find a function which describes the velocity, v, of the
jumper's ankles as a function of time during the free fall by
drawing a free-body diagram, applying Newton's Second Law of
Motion, separating variables, and integrating.

2. Use v to find a function, s, which describes the height of
the jumper's ankles as a function of time during free fall.

3.   Draw a free-body diagram showing all forces acting on the jumper when the bungee chord is first engaged. Use it along with Newton's Second Law to write a differential equation describing the height of the jumper's ankles as a function of time, starting when the bungee chord is first engaged.

4.   What are the boundary conditions for your equation? That is, what is the time t, height s and velocity v at the point in time when the bungee chord first engages?

5.   Set your calculator in differential equations mode and graphically solve the differential equation for s. (See Chapter 7 of your TI-85 guidebook) for details about the differential equations mode).   Determine the time interval on which this solution is valid by considering when the free-body diagram in Procedure 3 is valid. Use the trace mode on your calculator to find this interval and record your answer.

6.   Determine the maximum force exerted on the jumper's ankles by the chord during the initial fall.

7.   **G1**   Print (see Appendix 4.1) a graph of the solution to the differential equation (s vs. t) on top of the graph for the free fall to obtain a graph of s vs. t for the initial fall. This may require some creativity since the differential equation mode and the function mode cannot operate simultaneously. You may want to use the StorePic option.   (See Appendix 10.3)

8.   How much longer should the rope be so that the six foot tall jumper gets only part of his head wet? (Warning: These calculations are only <u>approximations</u>; do not try this at home!)

Review your calculations from the beginning to determine which calculations will be affected by changing the rope length. Redo any calculations which depend on the rope length to determine if the jumper's head gets wet. (Note: the tip of his head may be submerged by at most one foot.) **G2** Print a graph of the solution to the differential equation, height vs. time, which shows that only his head gets wet.

9.   If you were the jumper, hopefully you would not place your faith in the accuracy of these calculations. What assumptions would you be skeptical of?

Checklist of calculator graph printouts to be handed in:
☐   **G1**   Print a graph of the solution to the differential equation (s vs. t) on top of the graph for the free fall to obtain a graph for the initial fall.
☐   **G2**   Print a graph of the solution to the differential equation, distance vs. time, which shows that only his head gets wet.

Objectives:
1.   Use a graphing calculator to determine if a function could represent a probability density function.
2.   Analyze improper integrals by using a graphing calculator to calculate numerical integral.

Technology:
     TI-82, TI-85, or TI-92 graphing calculator.

Prerequisites:
     Introduction to improper integrals.

Overview: A probability density function P(x) is a function that defines the probability of an event which is described by a range of values of the variable x.   A probability density function, P(x), must satisfy the following conditions:
1. $P(x) \geq 0$ for any x in the domain of P.
2.   The probability that the variable, x, lies between a and b is the area of the region bounded by y=P(x), x=a, x=b and y=0 (see Figure 1).

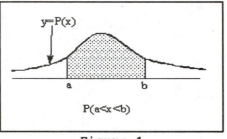

Figure 1

3.   The total area under the curve y=P(x) is 1.   That is, if the domain of P is (a,b), then $\int_a^b P(x)\ dx = 1$.

Procedures:

I.   Consider the function $f(x) = \dfrac{1}{\sqrt{2\pi}}\ e^{\frac{-x^2}{2}}$ where x is any real number.   The existence of the improper integral $\int_{-\infty}^{\infty} \dfrac{1}{\sqrt{2\pi}}\ e^{\frac{-x^2}{2}}\ dx$ will be investigated by analyzing $\lim\limits_{a \to -\infty} \int_a^0 f(x)\ dx$ and $\lim\limits_{b \to +\infty} \int_0^b f(x)\ dx$.

1.   Fill in Table 1 by calculating a numerical integral for $\int_a^0 f(x)\ dx$.   The format for $\int_a^b f(x)\ dx$ is fnInt (f(x),x,a,b).   See Appendix 8.4 for more details about fnInt.

a	$\displaystyle\int_{a}^{0} \frac{1}{\sqrt{2\pi}}\, e^{\frac{-x^2}{2}}\, dx$
-1	
-2	
-3	
-5	

Table 1

2.   Use these results to make a conjecture about $\displaystyle\lim_{a\to-\infty}\int_{a}^{0} \frac{1}{\sqrt{2\pi}}\, e^{\frac{-x^2}{2}}\, dx$.

3.   Fill in the following table by calculating a numerical integral for $\displaystyle\int_{0}^{b} \frac{1}{\sqrt{2\pi}}\, e^{\frac{-x^2}{2}}\, dx$.

b	$\displaystyle\int_{0}^{b} \frac{1}{\sqrt{2\pi}}\, e^{-\frac{x^2}{2}}\, dx$
1	
2	
3	
5	

Table 2

4.   Use these results to make a conjecture about $\displaystyle\lim_{b\to+\infty}\int_{0}^{b} f(x)\, dx$.

5.   Discuss the existence of $\displaystyle\int_{-\infty}^{\infty} \frac{1}{\sqrt{2\pi}}\, e^{-\frac{x^2}{2}}\, dx$ and give reasons for your conclusion.

6.   Discuss whether or not $f(x) = \dfrac{1}{\sqrt{2\pi}}\, e^{\frac{-x^2}{2}}$ satisfies conditions 1 and 3 in the definition of a probability density function.

II.   Consider the function $f(x) = \dfrac{2}{\sqrt{\pi}}\,\sqrt{x}\, e^{-x}$, $0<x<\infty$.   We will investigate the existence of the improper integral $\int_0^\infty \frac{2}{\sqrt{\Pi}} \sqrt{x}\, e^{-x}\, dx$

by considering $\lim\limits_{a \to +\infty} \int_0^a \frac{2}{\sqrt{\Pi}} \sqrt{x}\, e^{-x}\, dx$.

1. For each value of a in the following table, calculate a numerical integral for $\int_0^a \frac{2}{\sqrt{\Pi}} \sqrt{x}\, e^{-x}\, dx$ and record its value in the table.

a	$\int_0^a \frac{2}{\sqrt{\Pi}} \sqrt{x}\, e^{-x}\, dx$
1	
2	
3	
5	
10	

Table 3

2. Use these results to make a conjecture about $\lim\limits_{a \to +\infty} \int_0^a \frac{2}{\sqrt{\Pi}} \sqrt{x}\, e^{-x}\, dx$.

3. Discuss whether or not $f(x) = \frac{2}{\sqrt{\Pi}} \sqrt{x}\, e^{-x}$ could be a probability density function.

4. If your answer to question 3 is no but $\int_0^\infty \frac{2}{\sqrt{\Pi}} \sqrt{x}\, e^{-x}\, dx$ exists, how could you alter the function $f(x) = \frac{2}{\sqrt{\Pi}} \sqrt{x}\, e^{-x}$ to obtain a probability density function.

III. Consider the function $f(x) = \frac{2}{15\sqrt{\Pi}} x^{\frac{5}{2}} e^{-x}$, $0 < x < \infty$. We will investigate the existence of the improper integral $\int_0^\infty \frac{2}{15\sqrt{\Pi}} x^{\frac{5}{2}} e^{-x}\, dx$ by considering $\lim\limits_{a \to +\infty} \int_0^a \frac{2}{15\sqrt{\Pi}} x^{5/2} e^{-x}\, dx$.

1. Fill in the following table by calculating a numerical integral for $\displaystyle\int_0^a \frac{2}{15\sqrt{\pi}}\, x^{\frac{5}{2}}\, e^{-x}\, dx$.

a	$\displaystyle\int_0^a \frac{2}{15\sqrt{\pi}}\, x^{\frac{5}{2}}\, e^{-x}\, dx$
1	
3	
5	
10	
20	
50	

Table 4

2. Based on these results, could f(x) be a probability density function?

3. If your answer to question 2 is no, how could you alter

$f(x) = \dfrac{2}{15\sqrt{\pi}}\, x^{\frac{5}{2}}\, e^{-x}$ to obtain a probability density function.

IV. Consider the function $f(x) = 25.2\,(x^2)\,(1-x)^6$, $0<x<1$.

1. Calculate a numerical integral for $\displaystyle\int_0^1 25.2\,(x^2)\,(1-x)^6\, dx$.

2. Discuss whether or not $f(x) = 25.2\,(x^2)\,(1-x)^6$, $0<x<1$ could be a probability density function.

3. If your answer in question 2 is no, define a new function g in terms of f so that g is a probability density function.

V. Consider $f(x) = \dfrac{1}{\pi}\cdot\dfrac{1}{1+x^2}$, $-\infty<x<\infty$

1. Define the improper integral $\displaystyle\int_{-\infty}^{\infty} \frac{1}{\pi}\cdot\frac{1}{1+x^2}\, dx$ by using the definition in your textbook.

2. Find the indefinite integral $\displaystyle\int \frac{1}{\pi}\cdot\frac{1}{1+x^2}\, dx$.

3.   Analyze the improper integral $\int_{-\infty}^{\infty} \frac{1}{\pi} \cdot \frac{1}{1+x^2}\, dx$ by using the

Fundamental Theorem of Calculus and the definition from #1 above. Show all work in detail.

4.   Discuss whether or not f(x) could be a probability density function.

VI.   We have analyzed improper integrals both numerically (Procedures I-IV) and algebraically (Procedure V). We will now analyze improper integrals graphically.

1.   **G1**   Use the fnInt operation to print (see Appendix 4.1) a graph of the function F(x), where $F(x) = \int_{-x}^{x} \frac{1}{\pi} \cdot \frac{1}{1+t^2}\, dt$, for x>0.

2.   Discuss how you can tell if $f(t) = \frac{1}{\pi} \cdot \frac{1}{1+t^2}$ satisfies condition 3 in the overview by graphically analyzing F(x).

Checklist of calculator graph printouts to be handed in:

☐   **G1**   Use the fnInt operation to print a graph of the function F(x), where $F(x) = \int_{-x}^{x} \frac{1}{\pi} \cdot \frac{1}{1+t^2}\, dt$.

Objectives:
1.   Determine the effect of the parameters μ and σ on the graph of the normal probability density function.
2.   Determine the value of the probability, $P(a \leq x \leq b)$, by setting up an appropriate definite integral and using numerical techniques to approximate the integral.

Technology:
    TI-82, TI-85, or TI-92 graphing calculator.

Prerequisites:
    Introduction to definite integrals.

Overview: Many applied techniques of statistical decision making are based on the normal probability density function. This two parameter family of functions is defined by $f(x) = \dfrac{1}{\sqrt{2\pi}\,\sigma} e^{-\frac{(x-\mu)^2}{2\sigma^2}}$, where x is any real number. The primary purpose of this project is to investigate the roles of the two parameters, μ and σ. To study the effect of μ on the graph of f, we will fix σ and look for a pattern as we change the values of μ. In a similar manner, we will fix μ to study the effect of σ on the graph of f.

Procedures:
I.   For this procedure the parameter σ is assigned the value of 1. The role of μ will be explored by changing its value and observing the corresponding changes in the graph.
1.   **G1** Using a suitable viewing window, print out the graphs (see Appendix 4.1) of $y = \dfrac{1}{\sqrt{2\pi}} e^{-\frac{(x-\mu)^2}{2}}$ for $\mu=0$, $\mu=1$ and $\mu=3$ on the same screen. Label the graphs with "$\mu=0$", "$\mu=1$", and "$\mu=3$".

2.   What is the role of μ in describing the graphs of $f(x) = \dfrac{1}{\sqrt{2\pi}} e^{-\frac{(x-\mu)^2}{2}}$? Discuss translations and lines of symmetry.

3.   For each of the three curves trace to the point where $x=\mu$ and describe the significance of the point $(\mu, f(\mu))$.

4.   The parameter μ is called the mean of the probability distribution. Discuss why this name is appropriate.

II.   To study the role of σ, the parameter μ will be assigned a value of 0. Graphs of $f(x) = \dfrac{1}{\sqrt{2\pi}\,\sigma} e^{-\frac{x^2}{2\sigma^2}}$ will be observed for several values of σ to obtain information about the significance of σ.

1.   G2 Using a suitable viewing window, print out graphs of

$f(x) = \dfrac{1}{\sqrt{2\pi}\ \sigma}\ e^{-\frac{x^2}{2\sigma^2}}$ for σ=.5, σ=1, and σ=2 on the same screen. Label the graphs with "σ=.5", "σ=1", and "σ=2".

2.   What is the role of σ in describing the graphs of

$f(x) = \dfrac{1}{\sqrt{2\pi}\ \sigma}\ e^{-\frac{x^2}{2\sigma^2}}$?

3.   The parameter σ is called the standard deviation. Based upon the shapes of your graphs, discuss the statement: The standard deviation is a measure of the amount of deviation of x from the mean.

III. If the variable x has a normal distribution, then the probability that x lies between a and b, denoted by P(a<x<b), is

$\displaystyle\int_a^b \dfrac{1}{\sqrt{2\pi}\ \sigma}\ e^{-\frac{(x-\mu)^2}{2\sigma^2}}$.   P(a<x<b) may be interpreted as the percent of

all values of x which lie between a and b.

Normal distributions of the form $f(x) = \dfrac{1}{\sqrt{2\pi}\ \sigma}\ e^{-\frac{x^2}{2\sigma^2}}$, (with μ=0),

are used for all questions in this procedure.

1.   Find the percentage of x's which lie within one standard deviation of the mean, μ, when σ=.6. Since μ=0 we can write this symbolically   as   P(-σ<x-μ<σ)=P(-σ<x<σ)=P(-.6<x<.6)=

$\displaystyle\int_{-.6}^{.6} \dfrac{1}{\sqrt{2\pi}\ (.6)}\ e^{-\frac{x^2}{2(.36)}}\ dx.$ Use the operation fnInt to compute a

numerical approximation for the integral.   The format for

$\displaystyle\int_a^b f(x)\ dx$ is fnInt(f(x),x,a,b). See Appendix 8.4 for information

about fnInt.

2.   Find the percentage of x's which lie within one standard deviation of the mean, when σ=1.

3.   Find the percentage of x's which lie one standard deviation of the mean, when σ=2.

4.   Summarize your findings in questions 1-3.

5.   Find the percentage of x's that lie within 2 standard deviations of the mean for each distribution with μ=0 and σ=.6, σ=1 and σ=2.

6.   Summarize the results in question 5.

IV.   A math professor gave a 20 point math test to three different groups of students. An analysis of the results indicated that the scores for each of the three groups are approximately normally distributed. The mean for each group was 12. The standard deviations for Groups I, II and III were .5, 1 and 3 respectively. Letter grades were assigned by the following scale:

$$17 \le x \le 20 \quad A$$
$$14 \le x < 17 \quad B$$
$$9 \le x < 14 \quad C$$
$$5 \le x < 9 \quad D$$
$$x < 5 \quad F$$

1.   Let x represent the score of a student in Group I. The distribution for x is approximately $f(x) = \dfrac{1}{\sqrt{2\pi}\,(.5)}\, e^{-\frac{(x-12)^2}{2(.25)}}$. Find the approximate percentage of students who received A's, B's, C's, D's, and F's.

2.   Use the directions in question 1 as a guide to find the approximate percentage of students in Group II who received A's, B's, C's, D's, and F's.

3.   Use the directions in question 1 as a guide to find the approximate percentage of students in Group III who received A's, B's, C's, D's, and F's.

4.   Based on the results of these three questions, discuss the significance of the standard deviation, σ. You may find the words homogeneous and heterogeneous useful in describing a group.

V.   You are a member of one of three groups of students who take a 100 point test. Suppose that the scores for each group are approximately normally distributed and that the mean for each of the groups is 73. The standard deviations for groups I, II, and III are 5, 8, and 11, respectively. The instructor uses the following grade scale: $90 \le x \le 100$ A, $80 \le x < 90$ B, $70 \le x < 80$ C, $50 \le x < 70$ D, and $x < 50$ F.

1.   Suppose that you are a good student who hopes for an A. Tell which group you would prefer to be in and defend your answer using probabilities.

2.   Suppose that you are not at all concerned with your grade
in the class, provided that the grade is at least a C.   Tell
which group of students you would prefer to be in and defend your
answer using probabilities.

Checklist of calculator graph printouts to be handed in:

☐   **G1**   Print out the graphs of $y = \dfrac{1}{\sqrt{2\pi}} e^{-\frac{(x-\mu)^2}{2}}$ for $\mu=0$, $\mu=1$ and $\mu=3$ on the same screen.

☐   **G2**   Print out graphs of $y = \dfrac{1}{\sqrt{2\pi}\,\sigma} e^{-\frac{x^2}{2\sigma^2}}$ for $\sigma=.5$, $\sigma=1$, and $\sigma=2$ on the same screen.

Objectives:
1. Use the numerical integration feature of a graphing calculator to discover a formula for the mean, the standard deviation and the mode of a gamma distribution.
2. Explore relationships between a chi-square distribution and a normal distribution.

Technology:
 TI-82, TI-85, or TI-92 graphing calculator.

Prerequisites:
1. Introduction to improper integrals.
2. An introduction to factorial notation.
3. An introduction to the mean and the standard deviation of a distribution.

Overview: The mean or the median is usually used to describe the center of a distribution and the standard deviation is usually used to describe the spread of the data around the center. The mean and standard deviation of a probability distribution are defined in terms of the expected value, E(x), of a random variable x. E(x) is defined by $E(x) = \int_a^b x\, f(x)\ dx$ where (a,b) represents the range of values for the variable x. The mean μ and the standard deviation σ are defined as $\mu = E(x)$ and $\sigma = \sqrt{E(x^2) - (E(x))^2}$. The primary purpose of this project is to investigate the mean and standard deviation of a gamma distribution.

The gamma family of probability density functions is defined by $F(x) = \dfrac{1}{\alpha!\,\beta^{\alpha+1}}\ x^\alpha\ e^{-\frac{x}{\beta}}$ where $0 < x < \infty$ and $\beta > 0$. The values of α in this project are restricted to positive integers so that $\alpha!$ is an ordinary factorial.

Procedures:
I. The goal of this procedure is to discover a formula for the mean of a gamma distribution. Since a gamma function has two parameters, α and β, the formula for the mean will probably depend on both α and β.

It is generally more difficult to find a formula involving two variables than to find a formula involving only one variable. First we will restrict β and observe patterns connected to changes in α in order to conjecture a formula in terms of α. Next, we will observe patterns connected to changes in β and use these connections to change the first formula to one which depends on both α and β.

1. **G1** Print out the graphs (see Appendix 4.1) of four gamma functions where $\beta=1$ and α assumes the values of 1, 2, 3, and 6 on the same screen. Recall that the domain of a gamma function is $(0, +\infty)$. Label the graphs with "$\alpha=1$", "$\alpha=2$", "$\alpha=3$", and "$\alpha=6$".

2. Describe how the changes in α affect the mean of a gamma function.

3. Set up an improper integral which represents the mean for each of the gamma functions in question 1.

4. You should have found that when $\alpha=1$ and $\beta=1$, the mean for the gamma function is $\mu=\int_0^{+\infty} x(xe^{-x})\,dx = \int_0^{+\infty} x^2 e^{-x}\,dx$. Use the numerical integral feature, fnInt, to approximate $\int_0^a x^2 e^{-x}\,dx$ for a=1, a=10, a=20, a=50. The format for numerically approximating $\int_0^1 x^2 e^{-x}\,dx$ is fnInt $(x^2 e^{-x}, x, 0, 1)$. See Appendix 8.4 for details about fnInt. Record your results in Table 1.

5. Use the numerical integral feature, fnInt, and repeat question 4 with $\beta=1$ and $\alpha=2$, $\alpha=3$, and $\alpha=6$. Record your results in Table 1.

$\int_0^a \dfrac{x}{\alpha!\,(1)^{\alpha+1}}\, x^\alpha e^{-\frac{x}{1}}\,dx$				
a	$\alpha=1$	$\alpha=2$	$\alpha=3$	$\alpha=6$
1				
10				
20				
50				

Table 1

6. The numbers in the column headed $\alpha=1$ are values for $\int_0^1 x^2 e^{-x}\,dx$, $\int_0^{10} x^2 e^{-x}\,dx$, $\int_0^{20} x^2 e^{-x}\,dx$, and $\int_0^{50} x^2 e^{-x}\,dx$. Use these numbers to predict $\int_0^{+\infty} xe^{-x}\,dx$, which is the mean for the gamma distribution with $\beta=1$ and $\alpha=1$. Look for a pattern in the next column in Table 1 to predict the mean of the gamma distribution with $\beta=1$ and $\alpha=2$. Repeat this observation technique to determine the means of the gamma functions for $\alpha=3$ and $\alpha=6$. Fill in Table 2 with the means of the four gamma functions for $\beta=1$ and $\alpha=1,2,3$, and 6.

mean when $\beta=1$	$\alpha=1$	$\alpha=2$	$\alpha=3$	$\alpha=6$

Table 2

7.   Find a formula for the mean of a gamma function, with $\beta=1$, in terms of α by observing the pattern between values of α and the corresponding means in Table 2.

8.   The value of β has been restricted to 1 in order to determine a relationship between the mean and α. We will now let β vary to see how β affects the mean. Repeat the instructions in question 4 with the exception that $\beta=2$ rather than 1. Record your results in Table 3.

$$\int_0^a \frac{x}{\alpha!\,(2)^{\alpha+1}}\, x^\alpha e^{-\frac{x}{2}}\, dx$$

a	$\alpha=1$	$\alpha=2$	$\alpha=3$	$\alpha=6$
1				
10				
20				
50				

Table 3

9.   Use the results of Table 3 to predict the means of the four gamma functions for which $\beta=2$ and $\alpha=1$, $\alpha=2$, $\alpha=3$ and $\alpha=6$ and record your predictions in Table 4.

mean when $\beta=2$	$\alpha=1$	$\alpha=2$	$\alpha=3$	$\alpha=6$

Table 4

10.   Using the previous steps as a guide, find the means of four gamma functions with $\beta=3$ and $\alpha=1$, $\alpha=2$, $\alpha=3$, and $\alpha=6$. Record your results in the first row of Table 5. Fill in the other two rows with the means of the gamma functions from Tables 2 and 4.

	α=1	α=2	α=3	α=6
mean when β=3				
mean when β=2				
mean when β=1				

Table 5

11. How does changing β, from 1 to 2 and from 1 to 3, affect the means of the four gamma functions with α=1, α=2, α=3, and α=6?

12. Write a formula for the mean of a gamma function by adjusting the formula in question 7 to reflect the effect of β.

13. Use your formula to predict the mean of a gamma function with α=5 and β=4. Record your answer.

14. Check your answer by using the fnInt feature on your calculator to approximate the mean.

II. The goal of this procedure is to discover a formula for the standard deviation of a gamma function in terms of α and β. You will first find a formula for the variance σ^2 and use this formula to find a formula for the standard deviation σ.

$$\sigma^2 = E(x^2) - [E(x)]^2$$
$$\sigma = \sqrt{E(x^2) - [E(x)]^2}$$

1. Observe the four graphs from question 1 in Procedure I and determine which gamma function has the largest standard deviation and which one has the smallest standard deviation, based on how tightly clustered the data is.

2. To calculate σ^2, we need to first compute $E(x^2)$, which is

$$E(x^2) = \int_0^{+\infty} x^2 F(x) \ dx$$

$$= \int_0^{+\infty} \frac{x^2}{\alpha! \beta^{\alpha+1}} x^\alpha e^{-x/\beta} \ dx$$

For α=1 and β=1:

$$E(x^2) = \int_0^{+\infty} x^2 x e^{-x} \ dx$$

$$= \int_0^{+\infty} x^3 e^{-x} \ dx$$

Use the fnInt feature on your calculator to calculate $\int_0^{100} x^3 e^{-x} \ dx$.

Based on this value, predict the value of $\int_0^{+\infty} x^3 e^{-x}\, dx$, which is $E(x^2)$ and record your prediction.

3. Use this value of $E(x^2)$, along with the value of $E(x)$ (which is the same as the mean) for $\alpha=1$ and $\beta=1$ from Table 2, to predict σ^2, the variance of this gamma function.

4. Repeat questions 2 and 3 for gamma functions for which $\beta=1$, $\alpha=2$, $\alpha=3$, and $\alpha=6$. Fill in the Table 6 with the results from questions 2-5.

	$\beta=1$			
	$\alpha=1$	$\alpha=2$	$\alpha=3$	$\alpha=6$
$\int_0^{100} x^2 F(x)\, dx$				
$E(x^2)$				
$E(x)$				
σ^2				

Table 6

5. Use the results of the last row to predict a formula for the variance σ^2 in terms of α (when $\beta=1$).

6. Fill out Table 7 for a gamma function $F(x)$, with $\beta=2$ and $\alpha=1$, $\alpha=2$, $\alpha=3$, and $\alpha=6$.

	$\beta=2$			
	$\alpha=1$	$\alpha=2$	$\alpha=3$	$\alpha=6$
$\int_0^{100} x^2 F(x)\, dx$				
$E(x^2)$				
$E(x)$				
σ^2				

Table 7

7. Fill out Table 8 for a gamma function $F(x)$ with $\beta=3$ and $\alpha=1$, $\alpha=2$, $\alpha=3$ and $\alpha=6$.

$\beta=3$				
	$\alpha=1$	$\alpha=2$	$\alpha=3$	$\alpha=6$
$\int_{0}^{100} x^2 F(x)\ dx$				
$E(x^2)$				
$E(x)$				
σ^2				

Table 8

Analyze Tables 6-8 and notice how changing β from 1 to 2 to 3 affects the variance.

8.   Write a formula for the variance of a gamma distribution by adjusting the formula for variance when $\beta=1$ in question 5.

9.   Use this formula to write a formula for the standard deviation of a gamma distribution.

III.   The median of a gamma distribution $F(x)$ is a number m such that

$$\int_{0}^{m} F(x)\ dx = \int_{m}^{+\infty} F(x)\ dx = .5$$

These two integrals represent the areas depicted in Figure 1.

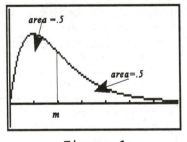

Figure 1

Use the numerical integral feature of a graphing calculator and the process of trial and error to estimate the median of the gamma function with $\alpha=4$ and $\beta=2$ accurate to 1 decimal place.

Checklist of calculator graph printouts to be handed in:
☐   **G1**   Print out the graphs of four gamma functions where $\beta=1$ and α assumes the values of 1, 2, 3, and 6 on the same screen.

Objectives:
1. Given a pair of polar coordinates, plot the graph of this point.
2. Transfer from rectangular coordinates of a point to polar coordinates.
3. Transfer from polar coordinates of a point to rectangular coordinates.
4. Given a polar name for a point, find two other polar names, one in which the first coordinate has a different sign and one whose second coordinate has a different sign.
5. Explore the points of intersection of two polar graphs.

Technology:
TI-82, TI-85, or TI-92 graphing calculator.

Prerequisites:
Good working knowledge of trigonometry with emphasis on special angles.

Overview:
The rectangular coordinate system is ideal for locating points, graphing functions, and finding areas of regions in most cases. However, the polar coordinate system is better for graphing some functions and finding areas when it is easier to subdivide a region into "wedges" rather than rectangles.

When a pair of numbers (a, b) is used to name a point in rectangular coordinates, this point is the intersection of a vertical line defined by x=a and a horizontal line defined by y=b. If a pair of numbers (a, b) represent a point in polar coordinates, the a and b are interpreted differently. The letter b defines a ray with its endpoint at the origin, which is called the pole. The measure of the angle between the polar axis (the positive x-axis) and the ray is b. The letter a indicates that the point (a, b) is |a| units

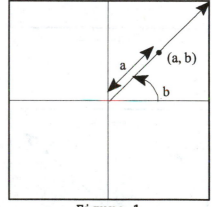

Figure 1

from the pole. Thus, the point (a, b) is the intersection of a ray defined by θ=b and a circle defined by r=a. See Figure 1.

Example 1: To graph the point (2, 1) in polar coordinates, visualize the pole and the polar axis. Next visualize a ray with endpoint O such that the measure of the angle AOB is 1 radian. Lastly, visualize a circle centered at the pole with a radius of 2 units. The intersection of the ray and the circle is the point (2, 1) as illustrated in Figure 2. Notice that we use the same notation to denote

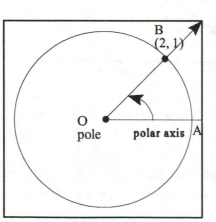

Figure 2

the ordered pair (2,1) as we do in rectangular coordinates. Therefore, the reader must determine from the context whether the point (2,1) means x=2 and y=1 (rectangular coordinates) or r=2 and θ=1 (polar coordinates).

Example 2:   A positive sign is assigned to the measure of an angle if the rotation from the polar axis is counterclockwise and a negative sign if the rotation from the polar axis is clockwise. To graph the point $(2, -\frac{\pi}{6})$, visualize the ray $\theta = -\frac{\pi}{6}$ and then the intersection of the ray and the circle r = 2.   See Figure 3.

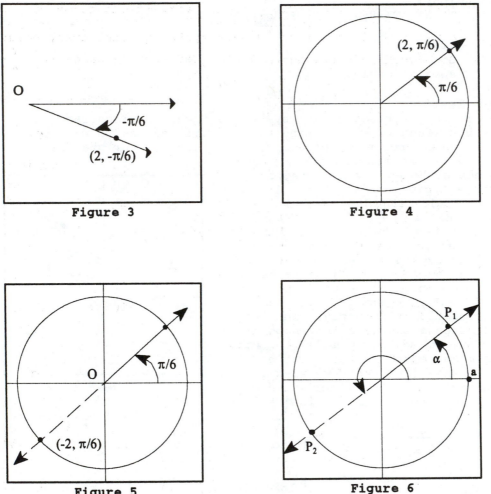

Figure 3

Figure 4

Figure 5

Figure 6

Examp
le 3: The graph of a polar pair (r, θ) when r < 0 involves a different interpretation. To graph the point $(2, \frac{\pi}{6})$ we will first consider r = 2. The condition r = 2 defines a circle centered at the pole with a radius of 2. Next consider the ray $\theta = \frac{\pi}{6}$. The intersection of the ray and the circle is the point

displayed in Figure 4.

To graph the point $(-2, \frac{\pi}{6})$ we will begin in a manner similar to graphing $(2, \frac{\pi}{6})$. Rather than finding the intersection of the circle and the ray $\theta = \frac{\pi}{6}$, find the point where the backward extension of the ray intersects the circle (see Figure 5).

A circle $r = a$ and a ray $\theta = \alpha$ define two points P_1 and P_2 as illustrated in Figure 6. Point P_1 is the intersection of the circle and the ray $\theta = \alpha$, and its coordinates are (a, α). P_2 is the intersection of the circle and the backward extension of the ray $\theta = \alpha$, and its coordinates are $(-a, \alpha)$. One of the disadvantages of the polar coordinate system is that every point has many names. Several names are listed for the point $(2, \frac{\pi}{4})$ in Figures 7-10.

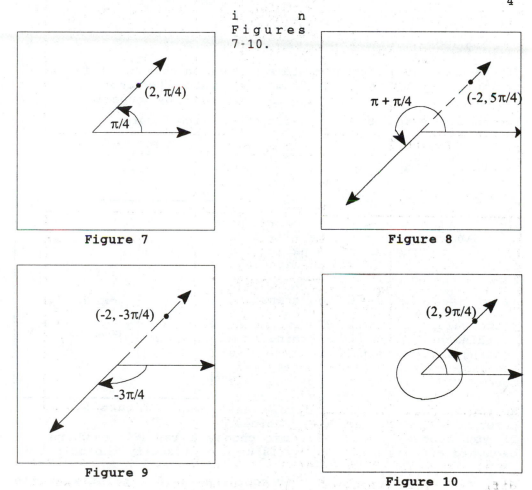

Figure 7	Figure 8
Figure 9	Figure 10

Procedures

I.

For each polar name, graph the point and find two other names, one for which the r coordinate has the opposite sign and one for which the θ coordinate has the opposite sign. Use the attached worksheet.

1. $(3, \frac{\pi}{6})$

2. $(-4, \frac{\pi}{3})$

3. $(2, -\frac{\pi}{4})$

4. $(-2, -\frac{\pi}{6})$

5. $(2, \frac{5\pi}{6})$

6. $(-3, \frac{11\pi}{6})$

7. $(-2, \frac{4\pi}{3})$

If one uses two different systems to name an object, it is useful to be able to transfer from one system to another. The rectangular-polar connection is obtained by superimposing a rectangular system over a polar system. From Figure 11 $\sin \theta = \frac{y}{r}$

or $y = r\sin \theta$

$\cos \theta = \frac{x}{r}$ or $x = r\cos \theta$

$r^2 = x^2 + y^2$ and $\tan \theta = \frac{y}{x}$

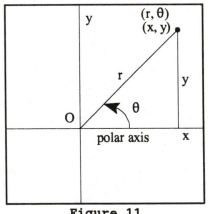

The transformation equations $y = r\sin \theta$ and $x = r\cos \theta$ are used to find the rectangular coordinates of a point when its polar coordinates are known. The equations $r^2 = x^2 + y^2$ and $\tan \theta = \frac{y}{x}$ are used to transform rectangular coordinates to polar coordinates. When transforming from rectangular coordinates, there are two choices for r and infinitely many for θ. However, the choices you make for r and θ must be paired appropriately. Draw a picture and refer to it as you make your choices.

Figure 11

If you choose r < 0, you must choose a ray θ = α where the backward extension of the ray is used to identify the point.

II. A. Find the rectangular coordinates of a point with the given polar coordinates and tell which quadrant the point is in.

1. $(2, \frac{\pi}{6})$

2. $(3, \frac{\pi}{2})$

3. $(-2, \dfrac{4\pi}{3})$

4. $(-3, \dfrac{11\pi}{6})$

B.   Find two sets of polar coordinates one with r>0 and one with r<0 for the point with the given rectangular coordinates.

1. $(2\sqrt{3}, 2)$

2. $(-4, -4)$

3. $(0, -3)$

4. $(\dfrac{1}{2}, 0)$

5. $(-6, 2\sqrt{3})$

__III.   Polar Mysteries__   Graph $r = -1 + \sin \theta$ and $r = 1$ in the same viewing window of your calculator (See Appendix 1.4 for details on polar graphing.)   How many points of intersection do you see?
To algebraically find the solution, we substitute $r = 1$ into $r = -1 + \sin \theta$ to get $1 = -1 + \sin \theta$ or $\sin \theta = 2$.   This equation has NO solution since $\sin \theta \leq 1$ for all θ.

What you see is not what you get!   Explain in detail why the "graphical solution" is not the same as this "algebraic solution".   Hint: Use the tracer to solve the mystery.

Worksheet:

1. $(3, \frac{\pi}{6})$

2. $(-4, \frac{\pi}{3})$

3. $(2, -\frac{\pi}{4})$

4. $(-2, -\frac{\pi}{6})$

5. $(2, \frac{5\pi}{6})$

6. $(-3, \frac{11\pi}{6})$

7. $(-2, \frac{4\pi}{3})$

Objectives:
1.   Given a name of a point in polar coordinates, find other polar coordinate names of that point. Names are to include pairs (r,θ) where $r<0$.
2.   Graph a polar equation, $r=f(\theta)$, and determine all values of θ, $0\le\theta\le2\pi$, for which $r<0$.
3.   Given a polar equation, find a second polar equation such that both graphs "look" the same but the points on the graphs are named differently.

Technology:
     TI-82, TI-85, or TI-92 graphing calculator.

Prerequisites:
1.   A good working knowledge of trigonometry and special angles.
2.   An introduction to polar coordinates.

Overview: Each point in a rectangular coordinate system has a unique name. This one to one correspondence between points and names does not exist in a polar coordinate system. In fact, a point in the plane has infinitely many polar names. For example, the point $\left(2,\frac{\pi}{6}\right)$ also has names of $\left(2,\frac{13\pi}{6}\right)$ and $\left(2,\frac{-11\pi}{6}\right)$. You should be able to list more names. The names are fairly obvious because there are many ways to name the ray $\theta=\frac{\pi}{6}$.

The interpretation of (r,θ) when r is a negative number is a little slippery. Consider the point whose polar coordinates are $\left(\frac{3}{2},\frac{\pi}{3}\right)$. To graph this point, first locate the ray $\theta=\frac{\pi}{3}$ and measure from the pole along the ray one and a half units. This point is the intersection of the ray $\theta=\frac{\pi}{3}$ and the circle $r=\frac{3}{2}$. See Figure 1.

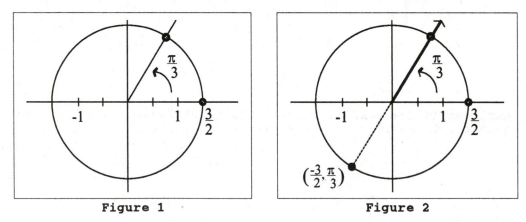

Figure 1   Figure 2

The point $\left(-\dfrac{3}{2},\dfrac{\pi}{3}\right)$ is the intersection of the backward extension of the ray $\theta=\dfrac{\pi}{3}$ and the circle whose radius is $\dfrac{3}{2}$. See Figure 2. Another name for this point is $\left(\dfrac{3}{2},\dfrac{4\pi}{3}\right)$. Notice that the ray $\theta=\dfrac{4\pi}{3}$ is the backward extension of the ray $\theta=\dfrac{\pi}{3}$.

The graph of a polar equation $r=f(\theta)$ is the set of all pairs (r,θ) when $r=f(\theta)$. Consider the function $r=-2+\cos(\theta)$. Since $\dfrac{3}{2}=-2+\cos\left(\dfrac{\pi}{3}\right)$, $\left(-\dfrac{3}{2},\dfrac{\pi}{3}\right)$ is on the graph of the function $r=-2+\cos(\theta)$.

Exploratory Exercise 1: For each ray, give the name of a ray which is its backward extension.

1. $\theta=\dfrac{\pi}{4}$ 4. $\theta=-\dfrac{\pi}{6}$

2. $\theta=\dfrac{5\pi}{6}$ 5. $\theta=\pi$

3. $\theta=\dfrac{4\pi}{3}$

Exploratory Exercise 2: Find a second polar name, with $r<0$, for each of the given points. A sketch is helpful.

1. $(2,\dfrac{\pi}{6})$ 5. $(3,-\dfrac{\pi}{3})$

2. $(4,\dfrac{5\pi}{3})$ 6. $(3,-\dfrac{5\pi}{6})$

3. $(3,\dfrac{3\pi}{4})$ 7. $(3,\dfrac{16\pi}{3})$

4. $(3,\dfrac{7\pi}{6})$

Exploratory Exercise 3: Find a second polar name, with $r>0$, for each of the given points. A sketch is helpful.

1. $(-2,\dfrac{\pi}{3})$ 4. $(-2,-\dfrac{\pi}{6})$

2. $(-2,\dfrac{5\pi}{6})$ 5. $(-2,-\dfrac{2\pi}{3})$

3. $(-2,\dfrac{4\pi}{3})$

Procedures:

I. Set the graphing mode of your calculator to Pol (polar) and the window format to PolarGC (see Appendix 1.4). Use the following window settings; $\theta_{min}=0$, $\theta_{max}=2\pi$, $\theta_{step}=\frac{\pi}{24}\approx.1308996$, $x_{min}=-3$, $x_{max}=3$, $y_{min}=-3$, and $y_{max}=3$. Graph each polar function in Table 1. Use the trace feature to determine all points between 0 and 2π for which $r<0$. Write your answers in column II by using interval notation. The endpoints of these intervals may be decimal approximations of the exact endpoints. Then, determine the exact end points of the intervals in column II and write your results in column III. There are several ways to determine the exact values of these endpoints. θstep is set at $\frac{\pi}{24}$ so that special angles such as $\frac{\pi}{6}$ are closely approximated by tracing. All of the endpoints for the functions listed in Table 1 will be special angles of the form $\frac{n\pi}{24}$, for some integer n. If you find $\theta=1.047198$, for example, then:

$$\frac{n\pi}{24}=1.047198$$

$$n=1.047798\cdot\frac{24}{\pi}=8$$

Hence, $\theta=\frac{8\pi}{24}=\frac{\pi}{3}$

You can check your guesses by determining if $r=0$ for this value of θ. You may pursue an algebraic approach by solving $r=f(\theta)=0$ for θ. These values of θ will be the endpoints of these intervals. If a part of a graph is duplicated over $[0, 2\pi]$, consider only the range of θ values which generate one complete graph.

I function	II Approximate Intervals	III Exact intervals
1. $r=2\cos\theta$		
2. $r=-2\cos\theta$		
3. $r=2\sin\theta$		
4. $r=-2\sin\theta$		
5. $r=1+\sin\theta$		
6. $r=1-2\sin\theta$		
7. $r=1-\cos\theta$		
8. $r=-1-\cos\theta$		
9. $r=-1+\sin\theta$		
10. $r=1-\sin\theta$		

Table 1

II. Graph $r_1=2+2\cos\theta$ and $r_2=-2+2\cos\theta$ on the same screen. How many graphs do you see? Go into the trace mode and use the down arrow key to switch between graphs. The number in the upper right corner of the screen indicates which function you are tracing. Using the trace mode, switch between graphs and observe what is happening. Describe why you only see one graph for two seemingly different functions. Explain in detail what is going on!

III. Graph each of the given functions. For each one, give a second function that has the same graph, but in which the corresponding points are named differently. Refer to Procedure II.

1. $r=5+3\sin\theta$ 5. $r=4-2\cos\theta$
2. $r=-4+2\sin\theta$ 6. $r=3+3\sin\theta$
3. $r=-3-\sin\theta$ 7. $r=\sin(2\theta)$
4. $r=-3-2\cos\theta$

Objectives:
1.   Estimate the coordinates of the points of intersection of two polar graphs by using the trace mode.
2.   Analytically find the coordinates of a point of intersection when the signs of the r coordinates are different for a point on two polar graphs.
3.   Use a calculator equation solver routine to estimate coordinates of points of intersection of two polar graphs.

Technology:
      TI-82, TI-85, or TI-92 graphing calculator.

Prerequisites:
      Good working knowledge of trigonometry with an emphasis on equation solving.

Overview: The task of finding all points of intersection of two polar graphs is sometimes tricky because a point has many polar names.   One approach is to set up and solve a trigonometric equation.   Unless one is clever, this technique does not produce the coordinates of all points of intersection.   Our goal is to utilize algebraic and graphical techniques to find or approximate a polar name for a point of intersection.   For the following procedures, set your calculator to the polar graphing mode and set the format so that polar coordinates are displayed in the trace mode (see Appendix 1.4).   Choose a window with θ from 0 to 2π and θstep = $\pi/24$.

Procedures:
I.   The goal of this procedure is to find the polar coordinates of points of intersection for the graphs of $r=\sin(2\theta)$ and $r=1$.

1.   First find the θ coordinate of all points whose name satisfies both equations by solving the trigonometric equation $\sin(2\theta)=1$ where $0 \leq \theta < 2\pi$.   Then find the value of r which corresponds to each value of θ and list all pairs in the form (r, θ).   Make sure that the pairs are solutions to both equations in their original form.

2.   Store and graph the equations $r=\sin(2\theta)$ and $r=1$ (see Appendix 1.4).   How many points of intersection do you see?

3.   Trace the graph of $r=\sin(2\theta)$ until you reach an apparent point of intersection and record the polar coordinates of this point in Table 1.   Change the cursor to the graph of $r=1$.   If the coordinates are approximately the same, record them in Table 1.   If they are different, trace on the graph of $r=1$ until you reach this point and record the coordinates in Table 1.

4.   Put the cursor on the graph of $r=\sin(2\theta)$ and trace until you reach a second point of intersection.   Record the coordinates of this point in Table 1.   Change the cursor to the graph of $r=1$.   If the cursor stays at approximately the same point, record the coordinates in Table 1.   If the cursor has moved to a different point on the graph of $r=1$, trace until you reach the second point of intersection again and record the coordinates of the point which belongs to the graph of $r=1$.

5.   Use problems 3 and 4 as a guide to record the coordinates

of the 3rd and 4th points of intersection by tracing r=sin(2θ)
and r=1. In Table 1 you should have two names (which may be the
same) for each point, one for r=sin(2θ) and one for r=1. What is
the relationship between the coordinates in columns 2 and 3 of
Table 1 for a point that is listed in your answer to problem 1?

point	name of the point on the graph of r=sin(2θ)	name of the point on the graph of r=1
first	(,)	(,)
second	(,)	(,)
third	(,)	(,)
fourth	(,)	(,)

Table 1

II.   The points which have the same name for each curve in Table
1 should correspond to those from question 1. If not, you have
probably failed to find all solutions of the equation in problem
1 of Procedure I. If the name of a point on the first graph is
different than the name of this point on the second graph, you
have not found a name which satisfies both r=sin(2θ) and r=1.
This point should not be in your initial list of solutions.
These are points of intersection of the two graphs that are not
found by solving the system of equations r=sin(2θ) and r=1
algebraically. The purpose of this procedure is to analyze the
results from Table 1 and use the observations to develop an
algebraic technique for finding the names of the points of
intersection which do not appear in list 1 of problem 1.

Select a point from Table 1 which has two different names.
Notice that the r coordinates have opposite signs and the θ
coordinate of one point is π more than the θ coordinate of the
other one. This relationship provides insight on how to change
the equation of a polar graph to find these points of
intersection algebraically.

Just as a point can have more than one polar name, a graph can be
defined by more than one polar equation.

In problem 1 of Procedure I we found solutions to the system:
        A1:r=sin(2θ)
        B1:r=1

To find a second equation for A1, replace θ by θ+π and r by -r.
        -r=sin[2(θ+π)]
        -r=sin[2θ+2π]
        -r=sin(2θ)
         r=-sin(2θ)

A second equation for the circle r=1 is r=-1. Now consider a
second system:
        A2:r=-sin(2θ)
        B2:r=-1

We will now look for other solutions by solving a system formed by pairing an equation from the second system with one in the first system.

1.   Find all solutions to the system containing A1 and B2 by solving a trigonometric equation.

2.   Find all solutions to the system containing A2 and B1 by solving a trigonometric equation.

3.   Find all solutions to the second system A2 and B2.

4.   Do each of the pairings A1 and B2, A2 and B1, and A2 and B2 produce "new" solutions that are different from those in problem 1 of Procedure I?   Explain why some pairings produce "new" solutions and why some do not.

III. Use the following plan to find the names of all points of intersection of two polar equations $r=f(\theta)$ and $r=g(\theta)$.
1.   Solve the equation $f(\theta)=g(\theta)$ and list all solutions in the form (r,θ).

2.   Graph both equations in the polar mode on the same screen.

3.   If you are <u>sure</u> that you have found the coordinates of all points of intersection in step 1 then you are done, and you may exit this plan.

4.   Use the trace to find a name for each point of intersection on each of the graphs.

5.   If you find a point of intersection such that the names are approximately the same on each graph, but the name is not in your list, you probably didn't find all solutions to the trigonometric equation in step 1.

6.   If there is a point of intersection such that you found a different name when you switched to the second graph, you have a point whose name is not in your list in #1. Use the technique outlined in Procedure II to change the name of one of the equations, pair them and solve until you have found all points of intersection of the two graphs.

7.   If you are not sure about the number of points of intersection, you should use the technique of Procedure II to help you decide.

8.   Check the pole, since there are situations in which the pole is a point of intersection, but it is not found by solving a system of polar equations.

For the following problems, find all points of intersection of the graphs of the polar equations. Show your work so that one can see that the names of all points are derived from solving a trigonometric equation.

1.   Find all points of intersection of the graph of r=1 and r=-1+cosθ.

2.   Find all points of intersection of the graph of r=-1+cosθ and r=cos(2θ).

3.   Find all points of intersection of the graphs of r=cos(2θ) and r=sinθ.

IV.   The purpose of this procedure is to estimate the coordinates of all points of intersection of the graphs of r=2sin(2θ)   and $r=\dfrac{4}{2-\cos\theta}$ for   0≤θ<2π.   The   equation $\dfrac{4}{2-\cos\theta}$=2sin(2θ) is rather difficult to solve algebraically. Since there is no root, solve or intersect routine in the polar graphing mode, we will use the rectangular coordinate mode to solve the system.

1.   Set your calculator in the function graphing mode and choose Xmin=0 and Xmax=2π. Store $y=\dfrac{4}{2-\cos x}$ and y=2sin(2x) on the y= menu and display their graphs on the same screen. Which variable is playing the role of θ? How many points of intersection do you see? Use your calculator to approximate the coordinates of the points of intersection and write your pairs in the form (r,θ). See Appendix 11.2 for details of solving a system of equations on the calculator.

2.   Go to the polar graphing mode and graph r=2sin(2θ) and $r=\dfrac{4}{2-\cos\theta}$ on the same screen. Choose θmin=0, θmax=2π and θstep=π/24. How many points of intersection are there?

3.   If there are points of intersection which are not in the list of problem 1, change the name of one of the equations and solve the new system. List all points of intersection in the form (r,θ).

Objectives:

1.   Interpret $\dfrac{dy}{dx}$ in polar coordinates.

2.   Interpret $\dfrac{dr}{d\theta}$ in polar coordinates.

Technology:
      TI-82, TI-85 or TI-92 graphing calculator.

Prerequisites:
1.   Interpret derivatives as slopes of tangent lines and as rates of change.
2.   Use derivatives to find local extrema.

Overview: Derivatives will be used to analyze graphs of functions of the form r=f(θ).   When in polar mode, your calculator can approximate two numerical derivatives, $\dfrac{dy}{dx}$ and $\dfrac{dr}{d\theta}$.   Choose the polar and radian modes on your calculator, and select polar GC (see Appendix 1.4).   Set up the window with ZoomTrig (see Appendix 2.2).   Check your window and notice that the range of θ is from 0 to 2π, where 2π ≈ 6.28 and θStep=$\dfrac{\pi}{24}$.

This value of θStep is chosen so that when you trace a graph, the θ values of points will include approximations for special angles such as $\dfrac{\pi}{6}$, $\dfrac{\pi}{4}$, and $\dfrac{\pi}{3}$.   You may have to adjust Xmin, Xmax, Ymin and Ymax depending on the graph. Read and work through Example 1 step by step to learn how to calculate the numerical derivatives $\dfrac{dy}{dx}$

and $\dfrac{dr}{d\theta}$.

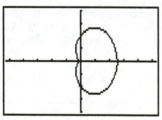

Figure 1

Example 1:   Store r=2+2cosθ at the polar function edit screen, display the graph, and compare your graph with the graph in Figure 1. Read Appendix 8.2 for more information about numerical derivatives. Approximate $\dfrac{dy}{dx}$ at the point (r,θ), where θ=$\dfrac{\pi}{3}$, by choosing $\dfrac{dy}{dx}$ on the appropriate menu, tracing to the point (r,θ) where θ=$\dfrac{\pi}{3}$ ≈ 1.047 and pressing **[Enter]**. The value of $\dfrac{dy}{dx}$ at θ=$\dfrac{\pi}{3}$ is approximately   zero.   Your   calculator   will   display   an approximation for zero such as -7.246×10^{-8} or 5.10955×10^{-12}.   In a similar manner, approximate $\dfrac{dr}{d\theta}$ at θ=$\dfrac{\pi}{3}$ (see Appendix 8.3).   Your result should be -1.732051.   If you get a different result, repeat the process.

Display the graph of r=2+2cosθ and use your calculator to draw a tangent line at the point (r,θ) where θ=$\dfrac{\pi}{3}$ (see Appendix 8.6.)

The tangent line appears to be horizontal and $\dfrac{dy}{dx}\Big|_{\theta=\pi/3}=0$ is the slope of this line. This should provide some insight on how to interpret $\dfrac{dy}{dx}$ at a point on a polar graph. The change in r with respect to θ, $\dfrac{dr}{d\theta}$, is about -1.73 at $\theta=\dfrac{\pi}{3}$. Trace the curve for values of θ from $\theta=0.39$ to $\theta=1.31$, and approximate $\dfrac{dr}{d\theta}$ at several points in this interval. Describe the behavior of r as θ increases over this interval, and relate this behavior to the sign of $\dfrac{dr}{d\theta}$ on the interval.

Procedures

I. For each function in 1-4 below, sketch its polar graph. Approximate the polar coordinates of all points on the graph for which a horizontal tangent line exists and $0\le\theta\le\pi$. Repeat the process for vertical tangent lines. Estimate the numerical derivatives $\dfrac{dy}{dx}$ and $\dfrac{dr}{d\theta}$ at these points. Use Worksheets Ia and Ib to organize your results.

1.   r=3cos(2θ)
2.   r=2+2sinθ
3.   r=2-2cosθ
4.   r=3cosθ

Worksheet Ia

Function	Approximate polar coordinates of points with a horizontal tangent line.	Value of $\frac{dy}{dx}$ at these points.	Value of $\frac{dr}{d\theta}$ at these points.
r=3cos 2θ			
r=2+2sinθ			
r=2-2cosθ			
r=3cosθ			

Worksheet Ib

Function	Approximate polar coordinates of points with a vertical tangent line.	Value of $\frac{dy}{dx}$ at these points.	Value of $\frac{dr}{d\theta}$ at these points.
r=3cos 2θ			
r=2+2sinθ			
r=2-2cosθ			
r=3cosθ			

Summarize your results, keeping in mind that the values for $\frac{dy}{dx}$ are only approximations. Is there a pattern for $\frac{dy}{dx}$ or $\frac{dr}{d\theta}$ at these points?

II: For each of the following functions:
a)   Sketch its polar graph.
b)   Estimate the polar coordinates of each point (r,θ) where 0≤θ≤π and r is a local maximum. Check by calculating the appropriate numerical derivatives. Use Worksheet II to organize your results.

Worksheet II

Function	Polar coordinates of points at which the radius is a local max.	Your check. Give appropriate values of $\frac{dy}{dx}$ or $\frac{dr}{d\theta}$, but not both.
r=3cos 2θ		
r=2+2sinθ		
r=2-2cosθ		
r=3cosθ		

Use your results in Procedures I and II to answer the following questions.

1.   Tell how and why you would use $\frac{dy}{dx}$.

2.   Tell how and why you would use $\frac{dr}{d\theta}$.

3.   Explain how you could use $\frac{dy}{dx}$ or $\frac{dr}{d\theta}$ to find local maximum values for r.

Objectives:
1. Graph and investigate the parameters of the families described by:

$$r = a\cos\theta$$
$$r = a\sin\theta$$
$$r = \sin(a\theta)$$
$$r = \cos(a\theta)$$
$$r = 2 + a\sin\theta$$
$$r = 2 + a\cos\theta$$
$$r = \frac{a}{1 + \cos\theta}$$
$$r = \frac{a}{1 - \cos\theta}$$

2. Using the results of the investigation, develop a family name for a particular set of curves.

Technology:
   TI-82, TI-85, or TI-92 graphing calculator

Prerequisites:
1. A good working knowledge of trigonometry.
2. An introduction to polar graphing.

Overview: In the rectangular coordinate system an equation of the form $y = ax^2 + bx + c$ where $a \neq 0$ defines a family of parabolas. This family of parabolas is sub-divided into types by placing restrictions on the parameter a. For example, a>0 defines a type of parabola which is concave up.

The purpose of this project is to investigate similarities between graphs of certain polar functions and group them into families. You should be able to visualize the graphs of certain families of functions for certain values of the parameters.

Example 1: Consider the polar equation defined by $r = 2\cos\theta$. Set your calculator for the polar (Pol) mode and (Radian) mode (see Appendix 1.4). Use the window setting:

$$\theta_{min} = 0 \qquad \theta_{max} = 2\pi \qquad \theta step = \frac{\pi}{24}$$

$$x_{min} = -2 \qquad x_{max} = 2$$
$$y_{min} = -2 \qquad y_{max} = 2$$

You may need to make appropriate window changes for some of the graphs that follow. The graph of $r = 2\cos\theta$ with these window settings is given in Figure 1. This is a distorted view as the graph appears to be elliptical. Zoom square (Zsquare) produces

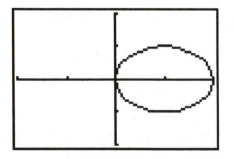

Figure 1

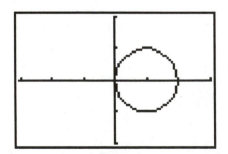

Figure 2

the graph in Figure 2. ZSquare redefines the window settings to approximately the same ratio (3/2) as the ratio of the number of horizontal pixels to the number of vertical pixels. This setting gives you the undistorted view in Figure 2. Throughout this project, use Zsquare to obtain a realistic view of the graph. The graph of r=2cosθ is a circle centered at (1,0) with a radius of one.   The polar equation r=acosθ represents a family of circles which pass through the pole and whose centers are on the polar axis.   The family consists of two types of circles depending on the sign of a.

Procedures: For each family of curves, choose appropriate values of the parameter a, graph the resulting function, and fill in the corresponding tables.

1.   r=asinθ

a	Describe the graph
1	
2	
-2	
-1.5	

Describe how the value of the parameter a affects the graphs.

2.   r=2sin(aθ) where a is an integer is called the family of roses.

a	Describe the graph
1	
2	
3	
4	
5	
-2	
-3	

a)   Describe how the value of a determines the type of rose that you get.

b)   Consider r=2sin(aθ) and r=2sin(bθ) where b=-a. Do these two types of equations produce different roses? Explain their similarities and their differences.

3.   r=cos(aθ)

a	Describe the graph
-2	
-1	
2	
3	

Describe how the value of the parameter a affects the graphs. How is this rose family like the rose family in problem 2? How does this rose family differ from the rose family in problem 2?

4. $r = 2 + a\sin\theta$

a	Describe the graph
$a < -2$	
$a = -2$	
$-2 < a < 0$	
$0 < a < 2$	
$a = 2$	
$a > 2$	

Describe how the value of the parameter a affects the graphs.

5.   Consider the function $r = 2 + a\cos\theta$. Refer to problem 4. Construct an appropriate table to summarize and describe how the value of the parameter a affects the graph.

Compare this family with the family in problem 4. Describe their similarities and their differences.

6.   Consider the function $r = \dfrac{a}{1 + \cos\theta}$. Design and fill in an appropriate table to determine a family name and analyze the different types of curves in this family.

What is an appropriate name for this family? How many different types does this family contain? Describe how the value of the parameter a affects the graphs.

7.   The graphs in problem 6 are symmetric with respect to the polar axis. Give a family of functions using one parameter that is identical to the one in problem 6, except that the curves of this family are symmetric about the line $\theta = \frac{\pi}{2}$ (the y axis in rectangular coordinates).

Objectives:
1.   Investigate the parameter a for the polar equations
r=$\dfrac{a}{1+a\cos\theta}$ and r=$\dfrac{a}{1-a\cos\theta}$ and identify the significance of
the parameter values.
2.   Given a polar equation of the form $\dfrac{pa}{1+a\cos\theta}$, $\dfrac{pa}{1-a\cos\theta}$,
$\dfrac{pa}{1+a\sin\theta}$ or $\dfrac{pa}{1-a\sin\theta}$, identify the conic and use a graphing
utility to verify your results.

Technology:
TI-82, TI-85 or TI-92 graphing calculator.

Prerequisites:
1.   A good working knowledge of trigonometry.
2.   An introduction to polar graphing (PC2).
3.   An introduction to conic sections.

Procedures: Set your calculator for the polar graphing mode (see Appendix 1.4). You will need to change the settings for x and y depending on the equation. The following settings are appropriate to start with:

$\theta_{min}=0$ $\theta_{max}=2\pi$ $\theta_{step}=.05$
$x_{min}=-1$ $x_{max}=1$ $x_{scl}=1$
$y_{min}=-1$ $y_{max}=1$ $y_{scl}=1$

I.   Consider the family r=$\dfrac{a}{1+a\cos\theta}$, a≥0. Graph the curve for
each value of the parameter a and fill in the table by describing each curve. If curve is a conic section, describe the major axis, transverse axis or axis of symmetry depending on the type of conic.

a	Describe r=$\dfrac{a}{1+a\cos\theta}$
.2	
.5	
.8	
1	
1.2	
1.5	
1.8	
2	

This family contains several types of conics. How many did you find? Name and describe each type and describe the role of a in determining the type of conic.

II.   Consider $r=\dfrac{a}{1-a\cos\theta}$, $a\geq 0$. Describe the similarities and differences between this family of curves and those in Procedure I.

III. Give a one parameter family of curves that is similar to the family in Procedure II, except the members of this family are symmetric with respect to $\theta=\dfrac{\pi}{2}$. Discuss the role of the parameter a.

IV. Write each of the following functions in the form $r=\dfrac{pa}{1 \pm a\cos\theta}$ or $r=\dfrac{pa}{1 \pm a\sin\theta}$ and give the values of p and a. Determine which type of conic section each equation represents.

1. $r=\dfrac{.4}{1-.4\cos\theta}$

2. $r=\dfrac{2}{1-\dfrac{2}{5}\cos\theta}$

3. $r=\dfrac{.6}{1-2\cos\theta}$

4. $r=\dfrac{-5}{5-2\cos\theta}$

5. $r=\dfrac{1}{1+\cos\theta}$

6. $r=\dfrac{1.6}{1+\cos\theta}$

7. $r=\dfrac{-.7}{1+\cos\theta}$

8. $r=\dfrac{2}{2+\sin\theta}$

9. $r=\dfrac{-1}{1+.5\sin\theta}$

10. $r=\dfrac{.1}{1+.5\sin\theta}$

V.

1.   What is the effect of p on the graph of a conic section $r=\dfrac{pa}{1\pm a\cos\theta}$ or $r=\dfrac{pa}{1\pm a\sin\theta}$ when compared with $r=\dfrac{a}{1\pm a\cos\theta}$ or $r=\dfrac{a}{1\pm a\sin\theta}$? Be complete. (You may need to look at additional conics in order to determine the significance of p.)

2.   What is the effect of a on the graph of a conic section?

Objectives:
1. Explore how graphs of r=f(θ) and r=af(bθ+c) are related.
2. Explore how two equations which look different can have graphs which look alike.

Technology:
TI-82, TI-85, or TI-92 graphing calculator.

Prerequisites:
1. Good working knowledge of trigonometry with emphasis on basic trigonometric identities.
2. Introduction to polar coordinates and graphs of cardioids and roses.

Overview:
Given the graph of a function y=f(x) in rectangular coordinates, you should be able to visualize and describe how the graph of y=f(x-a)+b is related to the given graph. The purpose of this project is to discover how the graphs of r=f(θ-c), r=af(θ) and r=f(bθ-c) are related to the graph of r=f(θ) for certain functions f.

Procedures:
Set your calculator for the polar graphing mode (see Appendix 1.4) and use the following window settings:

$$\theta_{min}=0 \qquad \theta_{max.}=2\pi \qquad \theta_{step}=\frac{\pi}{24}$$

You will probably need to change the settings for x and y depending on the equation. After you set up an appropriate window use ZOOMSQ to adjust the graphing window to display undistorted graphs.

I. Overlay the graph of the following polar equations on the graph of r=3cosθ. In each case describe how the graph is related to the graph of r=3cosθ. Set up an appropriate window and use ZOOMSQ to get a nondistorted view.

1. $r=3\cos(\theta-\frac{\pi}{4})$

2. $r=3\cos(\theta+\frac{\pi}{4})$

3. $r=3\cos(\theta+\frac{\pi}{2})$

4. $r=3\cos(\theta-\frac{\pi}{2})$

5. $r=3\cos(\theta-\frac{\pi}{6})$

Write a statement about the relationship between the graph of r=f(θ±c) and the graph of r=f(θ).

II. Graph each of the following pairs of polar equations and name the graph. Describe how the graph of the polar equation in column B is related to the graph of the equation in column A.

	A	B
1.	$f(\theta)=1-\cos(\theta)$	$g(\theta)=1-\cos(\theta-\frac{\pi}{4})$
2.	$f(\theta)=\sin(2\theta)$	$g(\theta)=\sin(2\theta-\frac{\pi}{4})$
3.	$f(\theta)=\sin(2\theta)$	$g(\theta)=\sin[2(\theta-\frac{\pi}{4})]$
4.	$f(\theta)=\cos(3\theta)$	$g(\theta)=\cos(3\theta-\frac{\pi}{2})$
5.	$f(\theta)=\cos(3\theta)$	$g(\theta)=\cos[3(\theta-\frac{\pi}{2})]$

Write a statement about the relationship between the graph of
$r=\sin(b\theta+c)$ and $r=\sin(b\theta)$. Write a statement about the
relationship between the graph of $r=\sin[b(\theta+c)]$ and $r=\sin(b\theta)$.
If you are unable to determine a relationship based on the above
graphs, graph other equations of the form $r=f(b\theta)$, $r=f(b\theta+c)$ and
$r=f[b(\theta+c)]$ where $f(\theta)=\cos\theta$ or $f(\theta)=\sin\theta$.

III. Graph each of the following pairs of polar equations.
Describe how the graph of the polar equation in column B is
related to the graph of the equation in column A.

A	B
1. $f(\theta)=\cos\theta$	$g(\theta)=\frac{1}{2}\cos\theta$
2. $f(\theta)=\cos\theta$	$g(\theta)=-2\cos\theta$
3. $f(\theta)=1-\cos\theta$	$g(\theta)=-2+2\cos\theta$
4. $f(\theta)=2+\cos\theta$	$g(\theta)=-4-2\cos\theta$
5. $f(\theta)=3+\cos\theta$	$g(\theta)=6+2\cos\theta$
6. $f(\theta)=1-\sin\theta$	$g(\theta)=-\frac{1}{2}+\frac{1}{2}\sin\theta$
7. $f(\theta)=1-2\sin\theta$	$g(\theta)=3-6\sin\theta$
8. $f(\theta)=-2+3\sin\theta$	$g(\theta)=2-3\sin\theta$
9. $f(\theta)=2-\sin\theta$	$g(\theta)=-1+\frac{1}{2}\sin\theta$
10. $f(\theta)=3-2\sin\theta$	$g(\theta)=1-\frac{2}{3}\sin\theta$

Write a statement about the relationship between the graphs of r=af(θ) and r=f(θ). Be careful about your conclusions.

Do the results generalize to functions such as $f(\theta)=\cos\left(\frac{3}{2}\theta\right)$ and $g(\theta)=-2\cos\left(\frac{3}{2}\theta\right)$?

IV.
1.   Determine if the graph of one of the following equations looks like the graph of one of the others. Tell which ones look alike. For each pair of equations whose graphs look alike, use the trace mode to determine if the corresponding points on the two graphs have the same name or different names. If the name of a point on one graph is different from the name of the same point on the other graph, describe how the names are related.

        a.   r=1+cosθ
        b.   r=1-cosθ
        c.   r=-1+cosθ
        d.   r=-1-cosθ

2.   Try to pick out pairs of equations whose graphs look the same without graphing. Use your observations from problem 1.

        a.   r=1-sinθ
        b.   r=-1+sinθ
        c.   r=1+sinθ
        d.   r=-1-sinθ

(Check your results by graphing.)

V.   A point has many polar names. For any point with coordinates (r,θ), the coordinates $(-r,\theta+\pi)$ give another name for this point. See Figure 1.

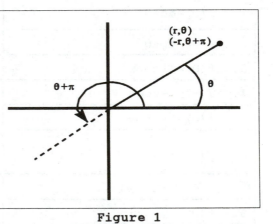

Figure 1

For each function below replace r by -r and θ by $\theta+\pi$ to obtain -r=f($\theta+\pi$). Write r as a function of θ and simplify if possible using appropriate trigonometric identities such as sin($\theta+\pi$)=-sinθ.

1.   r=1-sinθ
   a)   Does the graph of the new function look like the graph of r=1-sinθ?
   b)   Use the trace mode to compare the names of a point on both graphs. Does a point have the same name for each graph?

Use Problem 1 as a guide to answer questions a and b in problems 2-6.

2. $r=1+\cos\theta$

3. $r=\sin(2\theta)$

4. $r=\cos(2\theta)$

5. $r=\cos(3\theta)$

6. $r=1-2\cos\theta$

Summarize your observations in problems 1-6 about the graph of $-r=f(\theta+\pi)$ when compared with the graph of $r=f(\theta)$. This result does not generalize to all functions.

VI.

1.   Do the graphs the polar equations $r=\cos\left(\theta-\dfrac{\pi}{2}\right)$ and $r=\sin\theta$ look the same? Does a point have the same name for both graphs? Justify your answers by using trigonometry.

2.   For each of the following graphs, find an equation name in terms of the sine function and one in terms of the cosine function. The tick marks on both axes represent one unit. Check your answers by graphing the polar equations on your calculator.

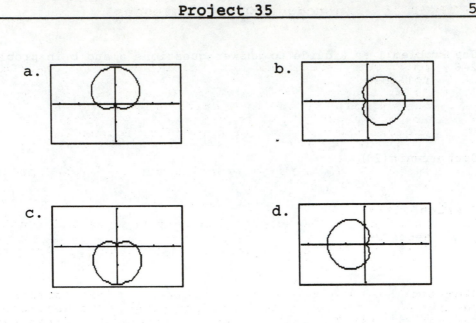

a.

b.

c.

d.

Objectives:
1. Introduce sequences as functions with restricted domains.
2. Calculate and list the terms of a given sequence with and without the use of a graphing calculator.
3. Graph a given number of terms of a sequence.
4. Explore the notion of a monotonic sequence.

Technology:
TI-82, TI-85 or TI-92 graphing calculator and program RECSEQ (for TI-85 only).

Prerequisites:
1. Knowledge of basic precalculus functions and function notation.
2. Introduction to limits of functions of the form $\lim_{x \to +\infty} f(x)$.

Overview: A sequence is a function whose domain is restricted to the set of positive integers. If a sequence is defined by $f(x)=3x+1$, then $f(1)=4$ is called the first term, $f(2)=7$, the second term, etc. In the symbol $f(1)$, 1 is the number of the term and $f(1)$ is the value of the first term. $f(\frac{1}{2})$ is not defined because the domain is restricted to the positive integers. To avoid notation confusion, n is used to represent the independent variable of a sequence and a_n is used to represent the dependent variable. We write $a_n=3n+1$ instead of $f(x)=3X+1$ to emphasize that the domain is restricted to positive integers. A sequence is a <u>discrete</u> function rather than a continuous one because there is a "gap" between any two numbers in the domain. The sequence command (see Appendix 9.1) on your calculator lists the values of the terms of a sequence. The command, $seq((-1)^x \frac{x}{x+1}, x, 1, 4, 1)$, displays $\{-.5, .6666666667, -.75, .8\}$ which is a list of the first four terms of the sequence $\{a_n\}$ defined as $a_n=(-1)^n \frac{n}{n+1}$. In the symbolism $seq((-1)^x \frac{x}{x+1}, x, 1, 4, 1)$, the first entry is the defining formula of the sequence, the second entry is the independent variable, the third is the number of the first term to be displayed, the fourth entry is the number of the last term, and the last entry is the increment for the variable x. Calculate the first four terms by hand and compare your results with those listed above. For example, the first term a_1 is obtained by replacing "n" with "1" in the formula $a_n=(-1)^n \frac{n}{n+1}$: $a_1=(-1)^1 \frac{1}{1+1}=-.5$

Procedures
I.   For each sequence in questions 1-3 below, find the value of the indicated terms using the seq command and hand calculate the values of those terms as a check. Record your results and write the particular sequence command and that you used.

1.   Find the first 3 terms of $a_n=\frac{n^2-1}{2n+1}$

2. a_{10}, a_{12}, and a_{14} for $a_n = \ln\left(\dfrac{n+3}{n+1}\right)$

3. a_{13}, a_{20}, and a_{27} for $a_n = \dfrac{n}{n+1}\cos(n\pi)$

Problems 4 and 5 are challenge problems. Calculate the value of the terms by hand.

4.   The first three terms of $a_n = \dfrac{2\cdot4\cdot6\cdots(2n)}{3\cdot5\cdot7\cdots(2n+1)}$

5.   The first four terms of $\{a_n\}$ where $a_1=3$ and $a_{n+1}=\dfrac{1}{2}a_n$

II.   The graph of the first six terms of the sequence defined by $a_n = \dfrac{2^n}{n^2}$ is given in Figure 1. The points on the graph of a sequence should not be connected because of the restricted domain.

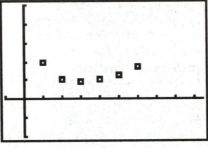

Figure 1

The tick marks represent 1 unit on both the x and y axis. Graph the first 6 terms of this sequence and compare the result with the graph in Figure 1. For details on sequence graphing see Appendix 9.3.

A sequence $\{a_n\}$ is increasing if $a_n \leq a_{n+1}$ for all $n\geq1$ or $a_1\leq a_2\leq a_3\leq\ldots$. The sequence $\{a_n\}$ is decreasing if $a_1\geq a_2\geq a_3\geq\ldots$ or if $a_n\geq a_{n+1}$ for all $n\geq1$. If a sequence is either increasing or decreasing it is called a monotonic sequence. Graph the first 50 terms of each of the given sequences. Use the tracer to determine if the sequence is monotonic for $1\leq n\leq50$ and if it is, tell if it is increasing or decreasing. If it is not monotonic for $1\leq n\leq50$, determine if it is monotonic for $m\leq n\leq50$ where $m\leq10$. Record your results.

1.   The first 50 terms of $a_n = \dfrac{n^2-1}{2n+3}$

2.   The first 50 terms of $a_n = \ln\left(\dfrac{n+3}{n+1}\right)$

3.   The first 50 terms of $a_n = \dfrac{n}{n+1}\cos(n\pi)$

4.   The first 50 terms of $a_n = \dfrac{5^n}{n!}$

5.   The first 50 terms of $a_n = \sqrt{n+1} - \sqrt{n}$

6.   The first 50 terms of $a_n = \left(\dfrac{3^n}{4^{n+1}}\right)\left(\dfrac{4^n}{3^{n+1}}\right)$

7.   The first 50 terms of $a_n = \dfrac{n!}{10^n}$

8.   The first 50 terms of $a_n = \ln\left(\dfrac{n^5}{n+1}\right)$

9.   The first 50 terms of $a_n = \dfrac{1}{n+\cos(n^2)}$

III.

1.   Consider the sequence whose first five terms are $a_1 = 1$, $a_2 = \dfrac{1}{2}$, $a_3 = \dfrac{1}{4}$, $a_4 = \dfrac{1}{8}$, $a_5 = \dfrac{1}{16}$.   Assuming that this trend continues, write the nth term for a_n in terms of n.

2.   The sequence $\{b_n\}$ is defined in terms of $\{a_n\}$ by $b_1 = a_1$, $b_2 = a_1 + a_2$, $b_3 = a_1 + a_2 + a_3$, etc. Use the sum command to fill in the entries in Table 1 with numbers in the form $\dfrac{m}{n}$, where m and n are integers. You can use the TI-82 or TI-92 to display the results as a fraction. For details about the sum command, see Appendix 9.2.

n	1	2	3	4	5	6
b_n						

Table 1

3.   Analyze this table to find a relationship between n and b_n.

4.   Use the formula in problem 3 to find b_9 and check this result by using the sum feature.

IV.   Two people A and B are playing a game where a pair of ordinary dice are rolled. The first person to roll the dice such that a "1" appears on each die is the winner. B politely tells A to roll first. There are 36 different outcomes for rolling a pair of dice so the probability that A will win on the first roll is 1/36. For A to win on his second roll, he must not win on his first roll and B must not win on his first roll. The probability that he will win on this second roll is $\frac{35}{36} \cdot \frac{35}{36} \cdot \frac{1}{36} = \left(\frac{35}{36}\right)^2 \frac{1}{36}$. The probability that he will win on his third roll is $\left(\frac{35}{36}\right)^4 \cdot \frac{1}{36}$.

1.   Write a formula for the sequence $\{A_n\}$ where A_n is the probability that A will win on his nth roll.

2.   Write a formula for the sequence $\{B_n\}$ where B_n is the probability that B will win on his nth roll.

3.   The probability that person A wins the game on or before the nth roll is $A_1 + A_2 + \ldots + A_n$. Find the probability that A will win the game on or before his 8th roll. Use the features of your calculator.

4.   Find the probability that person B will win the game on or before his 8th roll.

5.   If you were playing a game in which the winner received $100, would you let another person roll first? Tell why or why not.

Objectives:
1.   Find the value of a given term of a recursively defined sequence.
2.   Find the value of a given term by using the recursive features of a graphing calculator.
3.   Write an explicitly defined sequence as a recursive sequence.

Technology:
   TI-82, TI-85, or TI-92 graphing calculator and program RECSEQ (for TI-85 only)

Prerequisites:
   An introduction to compound interest.

Overview: An explicitly defined sequence is one whose n^{th} term is described in such a manner that you can find its value without evaluating any other terms of the sequence.   For example, the sequence $\{a_n\}$ whose defining formula is $a_n=3n-2$ is explicitly defined. We could easily find the value of a_{20} as $a_{20}=3(20)-2=58$. Consider a sequence $\{b_n\}$ where $b_1=1$ and $b_n=b_{n-1}+3$ for n>1. This sequence is recursively defined. To find b_{20}, one needs to find the value of b_1, b_2 all the way up to b_{19} . Sometimes a sequence is defined recursively rather than explicitly because of an application. The following example illustrates this point.

Example 1: $5,000 is invested for a period of four years. The interest will be compounded semi-annually with an annual rate of 7%, since 7% is an annual rate $\frac{7}{2}$% is the rate per compound period.   After six months, the amount of money P_1 is $5000+\left(\frac{.07}{2}\right)(5000) = 5000\left(1+\frac{.07}{2}\right)$. The principal will again change at the end of the second six months. Its value P_2 at this time is $P_2 = P_1+ \frac{.07}{2}P_1 = P_1\left(1+\frac{.07}{2}\right)$. After the third six month period, $P_3 = P_2+\frac{.07}{2}P_2$ or $P_3 = P_2\left(1+\frac{.07}{2}\right)$. In general $P_{n+1} = \left(1+\frac{.07}{2}\right)P_n$ where n is the number of compound periods. The value of the investment at the end of four years will be $P_8 = \left(1+\frac{.07}{2}\right)P_7$.

For this recursively defined sequence, the value of the 8th term depends on the value of the 7th which depends on the value of the 6th term, etc. In order to find the value of the 20th term one needs to calculate the value of the 19th term which depends on the 18th term, etc.

Procedures:
I.   Use the sequence graphing mode on a TI-82 or TI-92, or the program RECSEQ for a TI-85 to determine the value of the indicated term for the following sequences. For details about evaluating recursive sequences, see Appendix 9.4. Record your results.

1.   Given $a_1=2$ and $a_n=\sqrt{a_{n-1}}$, find a_{10}

2.   Given $a_1=1$ and $a_n=.7a_{n-1}$, find a_{13}

3.   Given $a_1=7$ and $a_n=.5\left(a_{n-1}+\frac{16}{a_{n-1}}\right)$, find a_{10}, a_{20}, a_{25}

4.   Given $a_1=3$ and $a_n=\left(1+\frac{1}{n^2}\right)a_{n-1}$, find a_6

5.   Given $a_1=1$ and $a_n=a_{n-1}+\frac{1}{n}$ find a_{15}

6.   Given $a_1=\frac{2}{3}$ and $a_n=a_{n-1}\frac{2n-2}{2n-1}$, find a_{15}

II. You initially invest \$2,000 at an annual rate of 8% compounded monthly. At the end of each month you invest an additional \$100.

Let P_1 be the amount of money that you have at the beginning of the first month.
$$P_1=2000$$
Let P_2 be the amount of money that you have at the beginning of the second month.
$$P_2=2000+2000\cdot\left(\frac{.08}{12}\right)+100=2000\left(1+\frac{.08}{12}\right)+100=P_1\left(1+\frac{.08}{12}\right)+100$$

1.   Write out a few more terms of the sequence $\{P_n\}$ and then write P_n recursively in terms of P_{n-1}.

2.   How much money will you have at the beginning of the 13th month?

3.   How long will it take for your money to double?

III. You have \$10,000 in a 'retirement' account which is set up with a 10% annual interest rate which is applied to the balance at the end of each month. After the interest is calculated you withdraw \$200.

Let P_1 be the amount of money at the beginning of the first month.
$$P_1=10,000$$

Let P_2 be the amount of money in the account at the beginning of the second month.
$$P_2=10000+\frac{.1}{12}(10000)-200=10000\left(1+\frac{.1}{12}\right)-200=P_1\left(1+\frac{.1}{12}\right)-200$$

1.   Write out a few more terms of the sequence $\{P_n\}$ and then write P_n recursively in terms of P_{n-1}.

2.   How many months can you receive a monthly payment of \$200 before the account is used up?

3.   How much will your last payment be?

4.   Suppose that at the beginning of the 15th month you decide to discontinue this plan and withdraw all of your money.

If there is no penalty for doing this, how much will you receive?

5.   Suppose that instead of withdrawing $200 each month, you want to withdraw the same amount at the end of each month so that this payment will continue for 48 months. Use a trial and error approach to find the amount of the monthly payment to the nearest dollar for the $10,000 'retirement' account.

IV.   You purchase a used car for its book value of $2200. Since you are broke, you put the full amount on a new credit card which has no other charges. You vow at this time not to charge anything else until the car is paid off. You have decided that you will pay $60 on the card at the end of each month. You check the fine print about the credit conditions and find that the annual interest rate on the unpaid balance is 18%.

1. P_1, the amount owed at the beginning of the first month, is $2200. P_2, the amount owed at the beginning of the second month is

$$P_2 = 2200 + .015(2200) - 60$$

Write P_n, the amount that you owe at the beginning of the nth month, recursively in terms of P_{n-1}.

2.   How many months will it take to payoff the credit card assuming that you continue to pay $60 per month until the last payment? How much is the last payment?

3.   Suppose that your car depreciates by 2% of its value each month. A_1, its value at the beginning of the first month, is $2200. A_2, its value at the beginning of the second month, is

$$A_2 = 2200 - (.02)(2200)$$
$$= 2200(1 - .02)$$
$$= 2200(.98)$$

Write the general term A_n for the value of this car at the beginning of the nth month.

4.   What is the value of the car when you make the last payment.

5.   Suppose you decide to sell this car for book value after you have owned it for 12 months and the book value is determined by the depreciation rule in question 2. If the selling price is less than what you owe, give the amount of money that you will have to come up with to pay off the credit card.

V.
1.   For the following sequences, write the given sequence recursively and then find the value of the 20th term. To do this it may be helpful to find a few terms and look for a pattern. Record your recursive definition and the value of a_{20}. See Appendices for information about recursive sequences.

a. $a_n = \dfrac{n!}{1 \cdot 3 \cdot 5 \cdots (2n-1)}$

b. $a_n = \dfrac{2 \cdot 4 \cdot 6 \cdots (2n)}{3 \cdot 5 \cdot 7 \cdots (2n+1)}$

c. $a_n = \dfrac{1 \cdot 3 \cdot 5 \cdots (2n-1)}{2^n n!}$

d. $a_n = \dfrac{1 \cdot 3 \cdot 5 \cdots (2n-1)}{2 \cdot 4 \cdot 6 \cdots (2n)}$

2.   Are the two sequences in problems c and d the same sequence? Explain your conclusion both graphically and algebraically.

Objectives:
1. Associate the limit of a sequence with a horizontal asymptote.
2. Use a graphing calculator to make a conjecture about the convergence or divergence of a sequence.
3. Describe the end behavior of a divergent sequence.
4. Write an explicitly defined sequence as a recursive sequence.

Technology:
   TI-82, TI-85, or TI-92 graphing calculator, and program RECSEQ (for the TI-85 only).

Prerequisites:
1. An introduction to horizontal asymptotes of a function.
2. An introduction to sequences.

The limit concept is used to describe the behavior of the terms of a sequence as the term number increases without bound. A sequence $\{a_n\}$ converges if its graph has a (unique) horizontal asymptote. If the equation of the horizontal asymptote is $y=c$, then the sequence converges to c. Symbolically, $\lim_{n\to+\infty} a_n = c$, means that the sequence $\{a_n\}$ converges to c. If the sequence's graph does not have a (unique) horizontal asymptote then the sequence diverges and it is said to be a divergent sequence.

The sequence graphing mode of a graphing calculator or program RECSEQ for TI-85 can be used to geometrically investigate convergence and divergence. In this mode one can control how many terms are graphed and what the beginning and ending terms are. See Appendix 9.3 for more details about graphing sequences.

Example 1: Consider the sequence defined by $a_n = \dfrac{n}{2n+1}$. Set your calculator for graphing sequences and the appropriate mode so that the points are not connected and graph the first 50 terms of $\{a_n\}$. Figure 1 shows a graph of the first 50 terms with the tracer activated.

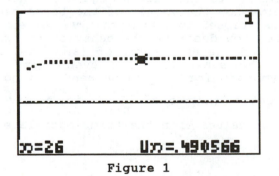

Figure 1

The values in Figure 1 indicate that $a_{26} = .49056603$. Tracing along the graph indicates that
$$a_{10} = .47619047$$
$$a_{20} = .48780487$$
$$a_{30} = .49180327$$
$$a_{40} = .49382716$$
$$a_{50} = .49504950$$

It appears that $\{a_n\}$ may converge to .5. We can further support this conjecture by graphing terms of the sequence for much larger values of n. Change your window settings so that n starts at 100000 and ends at 100050. By tracing this graph, we find that

$$a_{100000} = .49999750001$$
$$a_{100050} = .49999750126$$

If we are still not convinced that the limit of the sequence $\{a_n\}$ is .5, we can look at even larger values of n.

Example 2: The graph of the first 50 terms of the sequence defined by $a_n = \dfrac{2n^3 + 1}{n+3}$ is given in Figure 2.

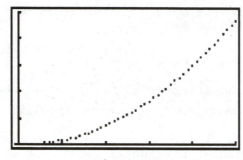

Figure 2

If you trace the graph, the terms of the sequence appear to increase without bound, and apparently there is no horizontal asymptote. You may provide more credibility to this conjecture by graphing with larger values of n. The symbols, $\lim\limits_{n \to +\infty} a_n = +\infty$,

indicate that the terms of the sequence $\{a_n\}$ increase without bound, and the sequence is divergent.

Example 3: Graph the first 50 terms of the sequence $\{a_n\}$ defined by $a_n = \dfrac{(-1)^n n}{n+1}$. With appropriate window settings, your graph should look similar to the one in Figure 3. Activate the trace mode, and observe how the values of the terms oscillate between positive numbers "close" to 1 and negative numbers "close" to -1. The tracer helps you describe the behavior of two subsequences, the odd terms, and the even terms of $\{a_n\}$. y=1 is a horizontal asymptote for the subsequence of even terms and y=-1 is a horizontal asymptote for the subsequence of odd terms. This

behavior indicates that $a_n = \dfrac{(-1)^n n}{n+1}$ is a divergent sequence

because for large values of n the terms oscillate between values

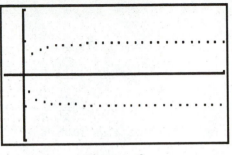

Figure 3

near -1 and 1.   In general, if a sequence contains two or more subsequences and each of the subsequences has a horizontal asymptote which is different from the other, then the original sequence diverges.

These three examples illustrate the difference between the end behavior of a convergent sequence vs. the end behavior of a divergent sequence.   This informal approach leads to reliable conjectures if you are careful in obtaining a "complete" graph of a sequence.

Procedures:
I.   Use the trace feature and the sequence graphing mode (RECSEQ for a TI-85) to determine if each of the given sequences is convergent or divergent.   If a sequence is convergent, give its limit.   If a sequence is divergent, describe its end behavior as increasing without bound, decreasing without bound, or oscillatory.

1. $a_n = \dfrac{1}{n}$ 2. $k_n = \dfrac{2n}{\sqrt[3]{n^3+1}}$

3. $b_n = \dfrac{1-(-1)^n}{2}$ 4. $l_n = \dfrac{n!}{n^2}$

5. $c_n = \dfrac{(-1)^n}{n}$ 6. $p_n = (1+n)^{\frac{1}{n}}$

7. $d_n = \dfrac{3n}{2n+1}$ 8. $p_n = \cos(n\pi)$

9. $f_n = \ln\left(\dfrac{n}{n+1}\right)$ 10. $q_n = \dfrac{\sin n}{n}$

11. $g_n = \dfrac{5^n}{n!}$ 12. $r_n = \dfrac{(-1)^n n}{n+1}$

II.   Each sequence in Procedure I is explicitly defined.   For example, the value of the 20th term of the sequence, $b_n = \dfrac{3n}{2n+1}$, is $b_{20} = \dfrac{3(20)}{2(20)+1} = \dfrac{60}{41}$.   The nth term of a recursive sequence is defined in terms of the previous term.   Consider the sequence $\{a_n\}$ where

$a_1=3$ and $a_n=a_{n-1}+\dfrac{1}{n}$.

$$a_2 \;=\; a_1+\frac{1}{2}=3+\frac{1}{2}$$

$$a_3 \;=\; a_2+\frac{1}{3}=3+\frac{1}{2}+\frac{1}{3}$$

$$a_4 \;=\; a_3+\frac{1}{4}=3+\frac{1}{2}+\frac{1}{3}+\frac{1}{4}$$

In order to find the values of the 20th term, one needs to find a_{19} which depends on a_{18}, etc.

The purpose of this procedure is to use the recursive graphing features of your calculator to investigate whether the given sequence is convergent or divergent. (See Appendix 9.4 for details on graphing recursive sequences).

If you think the sequence is convergent, give its limiting value. If you think that it is divergent, describe its end behavior as increasing without bound, decreasing without bound, or oscillatory.

1. $a_1=5$, $a_n=a_{n-1}+\dfrac{n}{10}a_{n-1}, n\geq 2$

2. $b_1=10$, $b_n=\dfrac{-.8n}{n+1}\,b_{n-1}$, $n\geq 2$

3. $c_1=5$, $c_n=.1\left(c_{n-1}+\dfrac{16}{c_{n-1}}\right)$, $n\geq 2$

III.   Sometimes it is advantageous to write an explicit sequence recursively. For example consider, $a_n=\dfrac{2\cdot 4\cdot 6\cdots (2n)}{3\cdot 6\cdot 9\cdots (3n)}$.

$$a_1 \;=\; \frac{2}{3}$$

$$a_2 \;=\; \frac{2}{3}\cdot\frac{4}{6}=a_1\cdot\frac{4}{6}=a_1\cdot\frac{2}{3}$$

$$a_3 \;=\; \frac{2\cdot 4\cdot 6}{3\cdot 6\cdot 9}=a_2\cdot\frac{6}{9}=a_2\cdot\frac{2}{3}$$

Recursively: $a_1=\dfrac{2}{3}$ and $a_n=\dfrac{2}{3}a_{n-1}$, $n\geq 2$. Write each of the given sequences recursively and record your recursive definition. Use the recursive graphing features of your calculator to investigate whether the given sequence is convergent or divergent. If the sequence converges, give its limit exactly. If it is divergent, describe its end behavior as increasing without bound, decreasing without bound, or oscillatory.

1. $a_n \;=\; \dfrac{n!}{1\cdot 3\cdot 5\cdots (2n-1)}$

2. $b_n \;=\; \dfrac{2\cdot 4\cdot 6\cdots (2n)}{3\cdot 5\cdot 7\cdots (2n+1)}$

3. $c_n = \dfrac{1\cdot3\cdot5\cdots(2n-1)}{2^n n!}$

4. $d_n = \dfrac{1\cdot3\cdot5\cdots(2n-1)}{2\cdot4\cdot6\cdots(2n)}$

5. $e_1 = \sqrt{2}$, $e_2 = \sqrt{2\sqrt{2}}$, $e_3 = \sqrt{2\sqrt{2\sqrt{2}}}$

Objectives:
1.   Use difference sequences to determine if a sequence can be explicitly named by a polynomial.
2.   Use difference sequences to determine the order of a polynomial sequence.
3.   Use polynomial regression to find an appropriate name for a polynomial sequence.

Technology:
     TI-82, TI-85 or TI-92 graphing calculator

Prerequisites:
     Introduction to sequences

Overview: When sequential data is collected, only the value of a finite number of terms is known. Sometimes the nature of the data leads to a recursive definition for the sequence. We will develop a method which uses difference sequences to find an explicit name for some of these sequences. A difference sequence for a given sequence $a_n = f(n)$ is defined as $\Delta a_n = f(n+1) - f(n) = a_{n+1} - a_n$. $\Delta^2 a_n$, the second difference sequence is defined as the difference of consecutive terms of the sequence Δa_n, $\Delta^2 a_n = \Delta a_{n+1} - \Delta a_n$. Likewise, $\Delta^3 a_n = \Delta^2 a_{n+1} - \Delta^2 a_n$.

Example 1: The purpose of this example is to show how to use difference sequences to gain information about an explicit formula for the nth term of a sequence. There is a property of difference sequences which is analogous to a property of derivatives. For example, if the 2nd derivative of a function is a non-zero constant, then one can conclude that the function is a 2nd degree polynomial. The analogous property states that if the second difference sequence is a non-zero constant, then one can conclude that the defining formula for the sequence is a second degree polynomial. To investigate this statement, consider a sequence a_n whose first five terms are: 4, 13, 28, 49 and 76. The diagram below illustrates how to organize the calculations to find the first four terms of Δa_k and the first three terms of $\Delta^2 a_n$.

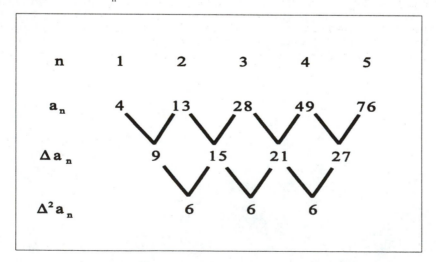

From this diagram, $\Delta a_1 = 9$, $\Delta a_2 = 15$, $\Delta a_3 = 21$, $\Delta a_4 = 27$. Also $\Delta^2 a_1 = 6$, $\Delta^2 a_2 = 6$ and $\Delta^2 a_3 = 6$. Note that the first three terms of the second difference sequence are the same. If this trend continues, we can then conclude that the defining formula for the sequence $\{a_n\}$ is a second degree polynomial. We will find a polynomial of the form $f(n)=an^2+bn+c$ which contains the points (1,4), (2,13), (3,28), (4,49) and (5,76). Use your calculator to perform a quadratic regression on the data. (See Appendix 7.3). The coefficients of the polynomial $f(n)=an^2+bn+c$ are a=3, b=0 and c=1. This is the "best" fit of a quadratic function to the five points. The results in Table 1 indicate that this polynomial sequence is a perfect fit for the five terms of $\{a_n\}$. Our conjecture looks like a good one! So the explicit name for the sequence is $a_n=3n^2+1$.

n	$f(n)=3n^2+1$
1	4
2	13
3	28
4	49
5	76

Table 1

This example illustrates that a constant second order difference sequence implies that the sequence itself is defined by a polynomial of degree 2. In general, a non-zero constant n^{th} order difference sequence implies that the original sequence is defined by a polynomial of degree n.

Example 2: Consider a sequence $\{a_n\}$ defined recursively by $a_1=1$ and $a_n=a_{n-1}+n^2$. Using this recursive definition, $a_1=1$, $a_2=1+4=5$, $a_3=5+3^2=14$ and $a_4=14+4^2=30$. We can evaluate any term of this sequence, but to find a_{50}, we need the value of a_{49}, and to find this value we need a_{48}, etc. In order to calculate a_{1000}, it is necessary to find the value of every term from the first to the 999th. To avoid such a hassle, we check out our conjecture about difference sequences to give us a hint on the degree of a polynomial which "fits" the sequence $\{a_n\}$. The first few terms of $\{a_n\}$ and the first and second difference sequences are listed in Table 2.

n	1	2	3	4	5	6	7
a_n	1	15	14	30	55	91	140
Δa_n	4	9	16	25	36	49	
$\Delta^2 a_n$	5	7	9	11	13		
$\Delta^3 a_n$	2	2	2	2			

Table 2

The third difference sequence $\Delta^2 a_n$ appears to be a constant sequence. We will conjecture that the sequence $\{a_n\}$ is a third degree polynomial. To check out the conjecture, a third degree polynomial is produced by using quadratic regression. Perform this regression. If you don't get $a_n = \frac{1}{3}n^3 + \frac{1}{2}n^2 + \frac{1}{n}n$, repeat the process (you may be a bit skeptical that this is the name of the sequence $\{a_n\}$ which is defined recursively by $a_1 = 1$ and $a_n = a_{n-1} + n^2$). Check this result by finding the value of a_8 using the recursive definition and also by using this formula. If you are still in doubt, you can check the result for values of $n > 8$.

Procedures: Use difference sequences to guess the degree of each polynomial sequence. Use the stat mode of your calculator to fit this data with a polynomial having this degree, as illustrated in Example 1 (see Appendices 7.2-7.5). Using these results, record the explicit name for each sequence.

1. $a_1 = 1$, $a_2 = 3$, $a_3 = 5$, $a_4 = 7$, $a_5 = 9$

2. $a_1 = 0$, $a_2 = 7$, $a_3 = 26$, $a_4 = 63$, $a_5 = 124$, $a_6 = 215$

3. $a_1 = -4$, $a_2 = 6$, $a_3 = 30$, $a_4 = 74$, $a_5 = 144$, $a_6 = 246$

4. $a_1 = 3$, $a_2 = 12$, $a_3 = 59$, $a_4 = 198$, $a_5 = 507$, $a_6 = 1088$, $a_7 = 2067$

5. $a_1 = -4$, $a_n = a_{n-1} + (3n - 5)$ for $n \geq 2$

Challenge Problem:
6.   The polynomial curve fitting features of a TI graphing calculator are limited to a polynomial of degree 4 or less. The following terms are from a polynomial sequence of degree 5. Set up an appropriate system of equations and then solve the system to determine the coefficients of the 5th degree polynomial. Research and use the appropriate features of your calculator to solve the system.

$a_1 = -2$, $a_2 = -37$, $a_3 = -128$, $a_4 = -179$, $a_5 = 146$, and $a_6 = 1543$

Objectives:
1.   Find an explicit name for a recursively defined sequence.
2.   Define and use a sequence to count the number of items in a geometrical configuration.
3.   Interpret n, a_n and s_n for items in a geometrical configuration.

Technology:
       TI-82, TI-85, or TI-92 graphing calculator.

Prerequisites:
       Introduction to sequences.

Procedures:
I.   Objects are placed to form the following triangular configurations.

1.   Draw the next configuration assuming that this trend continues.

2.   Consider the sequence $\{a_n\}$ where a_n is the number of elements in the nth row.

Row number:	Number of elements
n=1	$a_1=1$
n=2	$a_2=2$
n=3	$a_3=3$
n=4	$a_4=4$

What is a_7?

We will now define a sequence $\{s_n\}$ whose terms are the total number of elements in a triangular configuration with n rows.

Number of Rows in a Configuration	Total Number of Elements
1	$s_1=a_1=1$
2	$s_2=a_1+a_2=1+2=3$
3	$s_3=a_1+a_2+a_3=1+2+3=6$

The nth term, s_n, of this sequence is the sum of the first n terms of the sequence $\{a_n\}$. Symbolically, $s_n=a_1+a_2+a_3+\ldots+a_n$.

The sequence $\{s_n\}$ is called the sequence of partial sums of the sequence $\{a_n\}$.

3.   Find s_4, s_5, s_6, and s_7.

4.   Store integers 1-7 in lists L1 and s_1, s_2, . . . s_7 in list L2. Fit a polynomial curve to this data. Try polynomials of degrees 1 through 4 to find a perfect fit. See Appendices 7.2-7.5 for curve fitting information (If you are familiar with difference sequences, you can use them to determine the degree of the polynomial). Record your result in the form of an explicit name for s_n.

5.   How many items are in the 10th row of a triangular configuration that has at least 10 rows?

6.   Find the total number of items in a triangular configuration that has exactly 10 rows.

7.   Find the total number of items in a triangular configuration if the last row contains 200 items.

8.   A triangular configuration has a total of 5356 elements; how many rows are in this configuration?

9.   How many elements are in a section of a triangular configuration which still has 67 rows even after the first 8 rows have been deleted?

II. Round objects such as soccer balls are stacked to make a shape similar to a pyramid. Each layer is shaped like an equilateral triangle whose sides have one more object than the sides of the triangular layer directly above it. As you continue reading, try to visualize a "pyramid" stack. If a stack has a bottom layer which looks like this: •• The next and only layer above it has one item. Suppose a stack has a

bottom layer which contains six items.   The resulting pyramid
is shown in Figure 1.

Figure 1

1.   How many balls are in each triangular layer of the stack
in Figure 1?

2.   Fill in the second column of Table 1 where n is the layer
number and b_n is the number of balls in that layer.   Your
reasoning should produce a 7th layer with 28 balls.

layer number	number of balls in a layer	total # of balls in a stack of n layers
n	b_n	s_n
1	1	
2	3	
3		
4		
5		
6		
7		

Table 1

3.   Find an explicit formula for b_n.

4.   Consider the sequence $\{s_n\}$ where $s_n = b_1 + b_2 + \ldots + b_n$.   What do
the terms of $\{s_n\}$ represent?

5.   Fill in the third column of Table 1 with the values of s_n
for n=1-7.

6.   Find an explicit formula for s_n by using polynomial regression to find a perfect fit for the seven terms in Table 1. Try polynomials of degree 1-4 to fit the data points; only one of these curves will provide a perfect fit. Record your formula for s_n, writing the polynomial coefficients as fractions.

7.   Suppose that the bottom layer of a stack has 51 balls on a side. How many balls are in this layer?

8.   What is the total number of balls in this stack?

9.   What is the total number of balls in a stack if the bottom layer contains 55 balls?

10.   The third layer up from the bottom of a stack contains 210 balls. What is the total number of balls in the entire stack?

11.   A stack contains 5456 balls. How many layers are in this stack?

12.   There are 224 balls which are to be stacked in four separate stacks each containing the same number of balls. How many balls are on a side of the bottom triangular layer?

13.   How many balls are in the bottom layer?

14.   How is the sequence $\{s_n\}$ in Procedure I related to the sequence $\{b_n\}$ in Procedure II?

15.   How are the sequences $\{s_n\}$ in Procedures I and II related?

Objectives:
1.   Determine if a sequence is geometric.
2.   Describe the end behavior of a geometric sequence.
3.   Find an explicit name for a sequence of partial sums which is generated from a geometric sequence.
4.   Determine when the sequence of partial sums of a geometric sequence converges and, if it converges, find its limit.

Technology:
TI-82, TI-85 or TI-92 graphing calculator.

Prerequisites:
An introduction to sequences and sequences of partial sums.

Overview: A sequence $\{a_n\}$ is classified as geometric if $a_n=ra_{n-1}$ for all $n \geq 2$, where r is a constant. That is, each term of a geometric sequence is a constant multiple of the preceding term.

Example 1: The sequence $\{a_n\}$ defined by $a_1=3$ and $a_n=2a_{n-1}$ for $n \geq 2$, is called a recursive sequence because each term is defined in terms of the preceding term.

$a_1=3$
$a_2=2 \cdot a_1=2 \cdot 3$
$a_3=2 \cdot a_2=2 \cdot 2 \cdot 3=2^2 \cdot 3$
$a_4=2 \cdot a_3=2 \cdot 2 \cdot 2 \cdot 3=2^3 \cdot 3$
$a_5=2 \cdot a_4=2 \cdot 2 \cdot 2 \cdot 2 \cdot 3=2^4 \cdot 3$

The pattern above suggests that $a_n=2^{n-1} \cdot 3$ is an explicit formula for the nth term of the sequence $\{a_n\}$. An explicit formula is desirable because you can find the value of any term without calculating the value of any preceding terms. For example $a_{20}=2^{19} \cdot 3$.

Procedures:

I.   1.   Given $b_1=3$ and $b_n=\frac{1}{2} \cdot b_{n-1}$ for $n \geq 1$, fill out Table 1.

n	1	2	3	4	5	6	10	11	20
b_n	3	$\frac{3}{2}$	$\frac{3}{2^2}$						

Table 1

2.   Find an explicit formula for b_n.

3.   In the overview of this project we introduced a recursive definition for a geometric sequence. A geometric sequence, we said, must be of the form $c_1=a$ and $c_n=rc_{n-1}$, for $n \geq 2$. Use this recursive definition to find an explicit formula for a geometric sequence, c_n.

II.   For each sequence, $\{a_n\}$, tell if it is geometric and if it is, find a, r, and a_{20} (you do not have to simplify.)

1. $a_n=3n-1$

2. $a_n = \dfrac{2}{3^{n-1}}$

3. $a_n = n^2$

4. $a_n = 5\left(-\dfrac{1}{2}\right)^n$

5. $a_n = \left(-\dfrac{1}{2}\right)^n + \left(\dfrac{1}{2}\right)^n$

6. $a_1 = 3$, $a_2 = \dfrac{3}{5}$, $a_3 = \dfrac{3}{25}$, $a_4 = \dfrac{3}{125}$, . . .

7. $a_n = 3(1.2)^n$

8. $-\dfrac{1}{2}, 0, \dfrac{1}{2}, 1, \dfrac{3}{2}$, . . .

9. $-\dfrac{1}{2}, \dfrac{1}{4}, -\dfrac{1}{8}, \dfrac{1}{16}, -\dfrac{1}{32}$, . . .

10. $a_1 = 2$, $a_n = a_{n-1} + 4$, $n \geq 2$

11. $a_1 = 3$, $a_n = 2a_{n-1}$, $n \geq 2$

12. $a_1 = \sqrt{3}$, $a_2 = \sqrt[3]{3}$, $a_3 = \sqrt[4]{3}$ . . .

III.
1.   Consider a geometric sequence, $a_n = ar^{n-1}$. A few terms of
the sequence of partial sums, $\{s_n\}$, are
$$s_1 = a$$
$$s_2 = a + ar$$
$$s_3 = a + ar + ar^2$$
$$s_4 = a + ar + ar^2 + ar^3.$$
For each of the following geometric sequences write a recursive
definition for the sequence of partial sums, $\{s_n\}$. Use the

sequence graphing mode on a TI-82 or TI-92 (See Appendix 9.3) or the program RECSEQ for a TI-85 to look for trends in the end behavior of $\{s_n\}$ as n increases. Use these observations to determine if the sequence of partial sums is convergent or divergent. If $\{s_n\}$ converges, give its limit as a fraction.

a. $a_n=(-1)^n$

b. $a_n=4\left(\dfrac{1}{3}\right)^{n-1}$

c. $a_1=3$ and $a_n=(.1)a_{n-1}$ for $n\geq 2$

d. $a_n=(2)^n$

e. $-\dfrac{1}{2}, \dfrac{1}{4}, -\dfrac{1}{8}, \dfrac{1}{16}, -\dfrac{1}{32}$

f. $a_n=\dfrac{2}{3^{n-1}}$

g. $a_n=3\left(\dfrac{5}{4}\right)^{n-1}$

h. $a_1=2$ and $a_n=(1.2)\,a_{n-1}$ for $n\geq 2$

2.   Based on the results of problem 1, describe the conditions on a and r for the geometric sequence, $a_n=ar^{n-1}$, such that its sequence of partial sums converge.

IV.   A fellow student declares that it is easy to tell when the sequence of partial sums $\{s_n\}$ of a geometric sequence converges by observing r. He also states that when $\{s_n\}$ converges, its limit is $\dfrac{1}{r}$. In Table 2 you are given a and r for a geometric sequence, $a_n=ar^{n-1}$. Define $\{s_n\}$ recursively for each geometric sequence and use your calculator to find L, where $L=\lim\limits_{n\to+\infty} s_n$. You are to write L as a fraction.

a	r	Recursive def of s_n	Calculator estimate for $L = \lim\limits_{n \to +\infty} s_n$, (write as a fraction)	Student conjecture $\frac{1}{r}$
1	1/2			
1	1/3			
1	2/3			
1	1/4			
1	3/4			
1	1/5			
1	2/5			
1	3/5			

Table 2

Based on the results of this table, is the student's conjecture that $\lim\limits_{n \to +\infty} s_n = \frac{1}{r}$ correct? If you don't think so, write your own conjecture.

V. The conjectures in Procedure IV were based on geometric sequences restricted to a=1. We will now consider different values of a in order to determine the impact of a on the sum. Use your calculator to estimate L in the same way as in Procedure IV and fill in columns 3, 4, and 5 in Table 3.

a	r	Recursive def of s_n	Calculator estimate for L (Write as a fraction)	Your previous conjecture based on a=1
3	1/2			
5	1/3			
3	1/4			
3	1/5			
2	1/5			
1	2/3			

Table 3

Compare columns 4 and 5 to modify and record your final conjecture for L.

Objectives:
1. Introduce the concept of an infinite sum.
2. Explore the convergence or divergence of an infinite series.
3. Determine if a series is geometric and it if is, determine whether or not it is convergent.

Technology:
> TI-82, TI-85, or TI-92 graphing calculator, Program RECSEQ (for TI-85 only)

Prerequisites:
1. An introduction to sequences.
2. Concept of a limit.

Overview: The idea of "infinite" is a mystery to most people since their minds can only conceive of a finite number of tasks. The limit concept provides a means for analyzing an infinite process. The primary purpose of this project is to explore infinite sums. Based on the limit concept, we will use our finite thought processes to determine if an infinite sum makes "sense".

Consider a sequence $\{a_n\}$ and $s_n = a_1 + a_2 + \ldots + a_n$, its sequence of partial sums. The sequence $\{s_n\}$ consists of the sum of the first n terms of the sequence $\{a_n\}$, which is a finite sum. Since $\{a_n\}$ has an infinite number of terms, the sum of all of its terms is an infinite sum. This infinite sum is called a series and is symbolized by $\sum_{n=1}^{+\infty} a_n$. We study the behavior of the series by analyzing the corresponding sequence $\{s_n\}$. If the sequence of partial sums, $s_n = \sum_{n=1}^{n} a_n$, converges, then the infinite sum $\sum_{n=1}^{+\infty} a_n$, makes sense and is said to be convergent. If $\lim_{n \to +\infty} s_n = L$ then we assign the value L to the infinite sum, $\sum_{n=1}^{+\infty} a_n = L$. If $\lim_{n \to +\infty} s_n$ does not exist then the infinite sum $\sum_{n=1}^{+\infty} a_n$ doesn't make sense and is said to diverge.

You must be very careful when using your finite intuition to make a decision about an infinite process. Sometimes a finite mind will conclude that a statement is true when it is only a myth!

A common myth about an infinite series $\sum_{n=1}^{+\infty} a_n$ is:

If $\lim_{n\to+\infty} a_n = 0$ then $\sum_{n=1}^{+\infty} a_n$ converges. The following passage is an example of faulty reasoning to arrive at this conclusion.

> For large n, a_n is arbitrarily small. The sequence $\{s_n\}$ does not grow without bound since you are adding almost nothing to s_n to get s_{n+1}. Therefore $\{s_n\}$ converges and $\sum_{n=1}^{+\infty} a_n$ converges.

This is indeed a myth!

Procedures:

I.

1. Consider $s_n = 1 + \dfrac{1}{2} + \dfrac{1}{3} + \ldots + \dfrac{1}{n}$. Define s_n recursively as $s_1 = 1$ and $s_n = s_{n-1} + 1$ and use the sequence mode to store the recursive formula on a TI-82 or TI-92. (Use program RECSEQ for a TI-85). (See Appendix 9.4). Fill out the following table by writing the specified terms of s_n.

n	1	2	4	8	16	32	64
s_n	1	$1 + \dfrac{1}{2}$	$1 + \dfrac{1}{2} + \dfrac{1}{3} + \dfrac{1}{4}$				
calculated values of s_n	1	1.5	2.0833				

Based on the calculated values of the terms in this table, make a conjecture about the convergence or divergence of the sequence $\{s_n\}$.

2. Sometimes a sequence converges "slowly" or diverges "slowly" and you may be lead to an incorrect conjecture. Again consider the sequence $s_n = 1 + \dfrac{1}{2} + \dfrac{1}{3} + \ldots + \dfrac{1}{n}$, and suppose a cohort tells you that he thinks that $\lim_{n\to+\infty} s_n = 10$. Supply convincing evidence that this conjecture is false; be specific!

3. Study the following pattern carefully.

$$s_1 = 1$$

$$s_2 = 1 + \frac{1}{2} \geq \frac{3}{2}$$

$$s_4 = 1 + \frac{1}{2} + \frac{1}{3} + \frac{1}{4} \geq 1 + \frac{1}{2} + \frac{1}{4} + \frac{1}{4} \geq 1 + \frac{2}{2}$$

$$s_8 = 1 + \frac{1}{2} + \left(\frac{1}{3} + \frac{1}{4}\right) + \left(\frac{1}{5} + \frac{1}{6} + \frac{1}{7} + \frac{1}{8}\right) \geq 1 + 3 \cdot \frac{1}{2}$$

$$s_{16} = 1 + \frac{1}{2} + \left(\frac{1}{3} + \frac{1}{4}\right) + \left(\frac{1}{5} + \frac{1}{6} + \frac{1}{7} + \frac{1}{8}\right) + \left(\frac{1}{9} + \frac{1}{10} + \frac{1}{11} + \frac{1}{12} + \frac{1}{13} + \frac{1}{14} + \frac{1}{15} + \frac{1}{16}\right)$$

Therefore,

$$s_{16} \geq 1 + \frac{1}{2} + \frac{1}{2} + \frac{1}{2} + \frac{1}{2} \geq 1 + 4 \cdot \frac{1}{2}$$

Look for a pattern in the next four statements:

$$s_{2^1} \geq 1 + \frac{1}{2}, \quad s_4 = s_{2^2} \geq 1 + 2\cdot\frac{1}{2}, \quad s_8 = s_{2^3} \geq 1 + 3\cdot\frac{1}{2}, \quad s_{16} = s_{2^4} \geq 1 + 4\cdot\frac{1}{2}$$

a)   Find a value of k such that, $s_k \geq 64$
b)   Find a value of k such that, $s_k \geq 128$
c)   Find a value of k such that, $s_k \geq 500$
d)   Find a value of k such that, $s_k \geq 1000$
e)   Find a value of k such that, $s_k \geq 10,000$
f)   Find a value of k such that, $s_k \geq N$

Your above work indicates that $\{s_n\}$ grows without bound so $\lim_{n\to+\infty} s_n$ does not exist.

Therefore, the series $\sum_{n=1}^{+\infty} a_n$ diverges.

II. Explore each series below and determine if the series converges or diverges. Investigate $\lim_{n\to+\infty} s_n$ and attempt to determine whether or not this sequence grows without bound. Give a reason for each conclusion. To investigate the convergence of a series $\sum_{n=1}^{+\infty} a_n$, it is sometimes helpful to compare the finite sum $\sum_{k=1}^{n} a_k$ with $\sum_{k=1}^{n} \frac{1}{k}$ by writing the terms as:

$$1 + \frac{1}{2} + \frac{1}{3} + \ldots + \frac{1}{n}$$
$$a_1 + a_2 + a_3 + \ldots + a_n$$

Suppose that you notice that $a_1 \geq 1$, $a_2 \geq \frac{1}{2}$, $\ldots$ $a_n \geq \frac{1}{n}$. Since the finite sums associated with $\sum_{n=1}^{\infty} \frac{1}{n}$ grow without bound, this visual comparison may provide insight about the behavior of $\{s_n\}$ where $s_n = a_1 + a_2 + a_3 + \ldots + a_n$. One may conclude that $\sum_{n=1}^{+\infty} a_n$ diverges if $a_k \geq \frac{1}{k}$ for all k or if $a_k = \frac{c}{k}$ for all k.

1. $\sum_{n=1}^{+\infty} \frac{3}{n}$

2. $\sum_{n=1}^{+\infty} \left(2 + \frac{1}{n}\right)$

3. $\displaystyle\sum_{n=1}^{+\infty} \left(\frac{1}{n^2} + \frac{1}{n}\right)$

4. $\displaystyle\sum_{n=1}^{+\infty} \frac{1}{n+20}$

5. $\displaystyle\sum_{n=1}^{+\infty} \frac{1}{5n}$

6. $\displaystyle\sum_{n=1}^{+\infty} \left(\frac{n+1}{n}\right)$

There is a classification of sequences $\{a_n\}$ for which it is always easy to determine if $\displaystyle\sum_{n=1}^{+\infty} a_n$ converges or diverges. Consider the sequence $\{a_n\}$ defined by $a_n = ar^{n-1}$ and the associated series $\displaystyle\sum_{n=1}^{+\infty} (ar^{n-1})$. The sequences of partial sums, $\{s_n\}$, is $\displaystyle\sum_{n=1}^{n} ar^{n-1} = a + ar + ar^2 + \ldots + ar^{n-1}$. It is rather easy to show that $s_n = \dfrac{a}{1-r} - \dfrac{ar^n}{1-r}$. If $|r| < 1$ then $\displaystyle\lim_{n \to +\infty} \frac{ar^n}{1-r} = 0$ so $\displaystyle\lim_{n \to +\infty} s_n = \frac{a}{1-r}$. If $|r| \geq 1$ then $\displaystyle\lim_{n \to +\infty} s_n$ doesn't exist. Based on this information, $\displaystyle\sum_{n=1}^{+\infty} ar^{n-1}$ converges if $|r| < 1$ and $\dfrac{a}{1-r}$ is its value.

In this case $\displaystyle\sum_{n=1}^{+\infty} ar^{n-1}$ is associated with the number $\dfrac{a}{1-r}$ so the infinite sum makes "sense".

If $|r| \geq 1$ the series $\displaystyle\sum_{n=1}^{+\infty} ar^{n-1}$ diverges and this infinite sum doesn't make "sense". The series $\displaystyle\sum_{n=1}^{+\infty} ar^{n-1}$ is special because it is easy to find an explicit formula for s_n. For many series

$\sum\limits_{n=1}^{+\infty} a_n$ it is difficult to find an explicit formula for $\{s_n\}$ and

hence difficult to determine if $\sum\limits_{n=1}^{+\infty} a_n$ converges or diverges.

Even if you conclude that the series converges it is usually hard to determine what its limiting value is.

The following summary is provided to help you sort out information and use it to determine if a series converges.

1. $\sum\limits_{n=1}^{+\infty} a_n$ is an infinite series or an infinite sum.

2. $\{a_n\}$ is the generating sequence for the series.
3. $\{s_n\}$ is a sequence whose terms are a finite sum of the terms of $\{a_n\}$.

4.   Whether or not the infinite sum $\sum\limits_{n=1}^{+\infty} a_n$ makes sense depends

on the convergence or divergence of $\{s_n\}$.

III. For each series below, determine if the series is geometric or not. If so, determine whether the series is convergent or divergent. If the series is convergent, give its sum. If the series is not geometric, write its sequence of partial sums $\{s_n\}$ recursively. ($s_1 = a_1$ and $s_1 = s_{n-1} + a_n$). A table or a graph may be useful for exploring whether or not $\{s_n\}$ converges. Use the sequence mode on a TI-82 or TI-92 or the program RECSEQ for a TI-85.

1. $\sum\limits_{n=1}^{+\infty} \dfrac{2}{3^n}$

2. $\sum\limits_{n=1}^{+\infty} \dfrac{3^n}{5}$

3. $\sum\limits_{n=1}^{+\infty} \left(\dfrac{3}{5}\right)^n$

4. $\sum\limits_{n=1}^{\infty} n$

5. $\sum\limits_{n=1}^{\infty} 3\sqrt{n}$

6. $\sum\limits_{n=1}^{+\infty} 2$

7. $\displaystyle\sum_{n=1}^{+\infty} (.3)^{2n}$

8. $\displaystyle\sum_{n=1}^{+\infty} \frac{1}{n^2}$

9. $\displaystyle\sum_{n=1}^{+\infty} (-1)^n$

10. $\displaystyle\sum_{n=1}^{+\infty} \frac{1}{n}$

11. $\displaystyle\sum_{n=1}^{+\infty} \frac{100}{2^n}$

12. $\displaystyle\sum_{n=1}^{+\infty} \frac{1}{.3\sqrt{n}}$ (Hint: Use a table or a graph to compare terms

of this series with those of $\displaystyle\sum_{n=1}^{+\infty} \frac{1}{n}$)

13. $\displaystyle\sum_{n=1}^{+\infty} \frac{1}{0.2^n}$

Objectives:
1.   Given a function, find its approximating Taylor polynomial of degree n.
2.   Investigate trends in the error when using a Taylor polynomial to represent a function.

Technology:
     TI-82, TI-85, or TI-92 graphing calculator

Prerequisites:
     Differentiate a trigonometric, a logarithmic or an exponential function.

Overview: Your graphing calculator has the computing power to fit a polynomial to a set of data by using regression techniques. This method is limited to polynomials of degree 4 or less. The purpose of this project is to construct approximating polynomials of any degree and to investigate trends and approximation errors.

Example 1: The purpose of this example is to construct a polynomial that approximates sinx for x is close to zero. A first degree polynomial, $P_1(x)=a+bx$, has two parameters, a and b, so it is necessary to impose two conditions to determine a and b. Since we are interested in approximating $f(x)=sinx$ for numbers close to zero we will impose the condition that $P_1(0)=\sin(0)=0$. Since $P_1(0)=a$ we conclude that $a=0$. Next we will require that the slope of $P_1(x)$ at zero is the same as the slope of the tangent line to $f(x)=sinx$ at $(0,0)$.

Since $P_1'(x)=b$ and $f'(x)=cosx$, we conclude that:
$$b=P_1'(0)=f'(0)=\cos 0=1$$
So $P_1(x)=0+1x=x$.   The graphs of $P_1(x)=x$ and $f(x)=sinx$ are provided in Figure 1 for $-\frac{\pi}{2}\le x\le\frac{\pi}{2}$ and Figure 2 for $-\pi\le x\le\pi$.

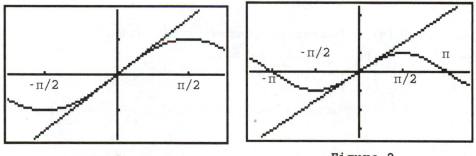

Figure 1

Figure 2

It is clear from these graphs that $P_1(x)=x$ is not a good estimate for $f(x)=sinx$ as $|x|$ increases. The graph of the error function $y=\left|sinx-P_1(x)\right|$ in Figure 3 provides a clear picture of the magnitude of error over the interval $\left[-\frac{\pi}{2},\frac{\pi}{2}\right]$.

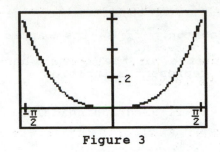

Figure 3

The maximum error over the interval $\left[-\dfrac{\pi}{2}, \dfrac{\pi}{2}\right]$ is $\left| P_1\left(\dfrac{\pi}{2}\right) - \sin\dfrac{\pi}{2} \right|$ =

$\dfrac{\pi}{2} - 1 \approx .57$.

Procedures:
I.   The next step is to construct a second degree polynomial $P_2(x)$ to approximate $f(x) = \sin x$ for values of x close to zero. Since a second degree polynomial, $P_2(x) = a + bx + cx^2$, has three parameters, we will impose the following three conditions on $P_2(x)$:

$$P_2(0) = f(0)$$
$$P_2{}'(0) = f'(0)$$
$$P_2{}''(0) = f''(0)$$

1.   Use these conditions to find a, b and c and write the resulting polynomial, $P_2(x)$.

2.   Is $P_2(x)$ a better approximator of $f(x) = \sin x$ than P_1 (x) over $\left[-\dfrac{\pi}{2}, \dfrac{\pi}{2}\right]$? Tell why or why not.

II.   Construct a third degree polynomial of the form $P_3(x) = a + bx + cx^2 + dx^3$ to approximate $f(x) = \sin x$, subject to the following conditions:

$$P_3(0) = f(0)$$
$$P_3{}'(0) = f'(0)$$
$$P_3{}''(0) = f''(0)$$
$$P_3{}'''(0) = f'''(0)$$

1.   Find a,b,c, and d and write the resulting polynomial, $P_3(x)$.

2.   Find the maximum error, $|\sin x - P_3(x)|$ over the interval $\left[-\dfrac{\pi}{2}, \dfrac{\pi}{2}\right]$ and record the result.

3.   Determine whether $P_2(x)$ or $P_3(x)$ is the better estimator for $y = \sin x$ over $\left[-\dfrac{\pi}{2}, \dfrac{\pi}{2}\right]$?

4.   What is the geometrical significance of each of the restrictions:
    a) $P_3(0) = f(0)$
    b) $P_3'(0) = f'(0)$
    c) $P_3''(0) = f''(0)$

III. We will now attempt to generalize a process for constructing polynomials, $P_n(x)$, to approximate a function $f(x)$ for values of x close to zero.   To find $P_3(x) = a + bx + cx^2 + dx^3$ for a function $f(x)$ the following conditions are imposed:
$$P_3(0) = f(0)$$
$$P_3'(0) = f'(0)$$
$$P_3''(0) = f''(0)$$
$$P_3'''(0) = f'''(0)$$

$P_3(x) = a + bx + cx^2 + dx^3$ so $P_3(0) = a = f(0)$
$P_3'(x) = b + 2cx + 3dx^2$ so $P_3'(0) = b = f'(0)$
$P_3''(x) = 2c + 6dx$ so $P_3''(0) = 2c = f''(0)$
$P_3'''(x) = 6d$ so $P_3'''(0) = 6d = f'''(0)$

Therefore $a = f(0)$, $b = f'(0)$, $c = \dfrac{1}{2}f''(0)$ and $d = \dfrac{1}{6}f'''(0)$,   so

$P_3(x) = f(0) + f'(0)x + \dfrac{1}{2}f''(0)x^2 + \dfrac{1}{6}f'''(0)x^3$.   An nth degree polynomial that is constructed with the following restrictions is called a Taylor polynomial:

$$P_n(0) = f(0), \quad P_n'(0) = f'(0), \quad P_n''(0) = f''(0) \quad \ldots \quad P_n^{(n)}(0) = f^{(n)}(0)$$

1.   Using $P_4(x) = a + bx + cx^2 + dx^3 + ex^4$, find a fourth degree Taylor polynomial for approximating a function $f(x)$ for x-values close to zero.   The parameters a-e are to be written in terms of $f(0)$, $f'(0)$, $f''(0)$, $\ldots$ $f^{(n)}(0)$.

2.   Complete Table 1 for Taylor polynomials.

n	Taylor Polynomials
1	$P_1(x) = f(0) + \dfrac{f'(0)x}{1!}$
2	$P_2(x) = f(0) + \dfrac{f'(0)x}{1!} + \dfrac{f''(0)x^2}{2!}$
3	$P_3(x) = f(0) + \dfrac{f'(0)x}{1!} + \dfrac{f''(0)x^2}{2!} + \dfrac{f'''(0)x^3}{3!}$ (Note $3! = 1 \cdot 2 \cdot 3 = 6$)
4	$P_4(x) =$
6	$P_6(x) =$
k	$P_k(x) =$

Table 1

IV. For a function $f(x)$, subject to the appropriate conditions, an nth degree Taylor polynomial $P_n(x)$ for approximating $f(x,)$ for values of x close to zero, is given by

$$P_n(x) = f(0) + \frac{f'(0)x}{1!} + \frac{f''(0)x^2}{2!} + \frac{f'''(0)x^3}{3!} + \ldots + \frac{f^{(n)}(0)x^n}{n!}.$$

Taylor polynomials $P_1(x)$, $P_2(x)$, $P_3(x)$ and $P_4(x)$ for $f(x) = \cos x$ about zero are:

$$P_1(x) = 1 + \frac{0x}{1!} = 1$$

$$P_2(x) = 1 + \frac{0 \cdot x}{1!} + \frac{1 \cdot x^2}{2!} = 1 + \frac{x^2}{2!}$$

$$P_3(x) = 1 + \frac{0 \cdot x}{1!} + \frac{1 \cdot x^2}{2!} + \frac{0 \cdot x^3}{3!} = 1 + \frac{x^2}{2!}$$

$$P_4(x) = 1 + \frac{0 \cdot x}{1!} + \frac{1 \cdot x^2}{2!} + \frac{0 \cdot x^3}{3!} + \frac{1 \cdot x^4}{4!} = 1 + \frac{x^2}{2!} + \frac{x^4}{4!}$$

1. Find Taylor polynomials of degrees 6 through 10 about zero for $f(x) = \cos x$ and record your results.

2. Why is a Taylor polynomial of degree $n = 2k$ the same as a

Taylor polynomial of degree n=2k+1 for f(x)=cosx about 0?

3. **G1** Printout the graphs (see Appendix 4.1) of polynomials $P_1(x)$, $P_2(x)$ and $P_4(x)$ overlaid on the graph of f(x)=cosx over the interval $\left[-\frac{\pi}{2}, \frac{\pi}{2}\right]$.

4. Discuss any error trends from this graph.

5. For $P_6(x)$ what is the error trend over an interval [-m,m] as m increases?

V.
1. Construct a Taylor polynomial of degree 7 for $f(x)=e^x$ about x=0 and record the results.

2. **G2** Printout the graphs of Taylor polynomials of degree 2, 4, and 7 for $f(x)=e^x$ overlaid on the graph of $f(x)=e^x$.

3. Discuss any error trends that you observe.

Checklist of calculator graph printouts to be handed in:

☐   **G1**   Graph of Taylor polynomials of degrees 1, 2, and 4 overlaid on the graph of f(x)=cosx over the interval $\left[-\frac{\pi}{2}, \frac{\pi}{2}\right]$.

☐   **G2**   Graph of Taylor polynomials of degrees 2, 4 and 7 for $f(x)=e^x$ overlaid on the graph of $f(x)=e^x$.

Objectives:
1.   Find a Taylor polynomial to approximate ln x for 0<x<2.
2.   Find a Taylor polynomial to approximate ln x for x≥2.

Technology:
      TI-82, TI-85 or TI-92 graphing calculator

Prerequisites:
1.   Introduction to Taylor Polynomials
2.   Ability to take higher order derivatives of $y=\ln\left(\dfrac{1+x}{1-x}\right)$

Overview: The purpose of this project is to use Taylor polynomials to approximate ln x, where x is a positive real number. It is not possible to construct a Taylor polynomial about x=0 since ln x and its derivatives are not defined at x=0. The procedures of this project are designed to explore alternative techniques for finding Taylor polynomials to approximate ln x.

Procedures:
I.
1.   Consider f(x)=ln(1+x) which is a translation of y=ln x. Find Taylor polynomials of degrees 3 and 5 about x=0 for f(x)=ln(1+x). Record your formulas for $P_3(x)$ and $P_5(x)$.

2.   **G1** Print the graphs (see Appendix 4.1) of $y=P_3(x)$ and $y=P_5(x)$ overlaid on the graph of f(x)=ln(1+x).

3.   The purpose of finding $P_5(x)$ is to use it as an approximation for f(x)=ln(1+x). Graph the error function $E_5(x)=|P_5(x)-\ln(1+x)|$. For what values of x does $P_5(x)$ provide an error less than .1?

4.   Use $P_5(x)$ to approximate ln(1+x) and estimate the error:
      a)   ln(1+-.5)

      b)   ln(1+.5)

      c)   ln(1+1.5)

5.   Can you find an integer n such that $P_n(1.5)$ approximates ln (1+1.5) with an error less than .1? If yes, give the value of n that you used. If not, give the largest n that you tried and estimate the error for this n.

II.   In this procedure we will find a Taylor polynomial to approximate ln 2.5 with an error less than .1.

1.   Consider the function $f(x) = \ln\left(\dfrac{1+x}{1-x}\right)$.   Find Taylor polynomials $P_3(x)$ and $P_5(x)$ about $x=0$ for $f(x)$.

2.   What value of x would you choose in order to use $P_5(x)$ as an approximation for ln 2.5?

3.   Use $P_5(x)$ to approximate ln 2.5.

4.   Estimate the error when $P_5(x)$ is used to approximate ln 2.5 by evaluating ln 2.5 and comparing this number to the result of question 3.   Compare this error with the result of question 4c in Procedure I and comment on your comparison.

5.   Estimate ln (10) by using $P_5(x)$.

Challenge Problem: It can be shown that $P_n(x)$ provides good approximations for $\ln\dfrac{1+x}{1-x}$ for $-1<x<1$.   Find the corresponding values of a such that $P_n(x)$ provides good approximations for ln a.

Checklist of calculator graph printouts to be handed in:
☐   **G1**   Print the graphs of $y=P_3(x)$ and $y=P_5(x)$ overlaid on the graph of $f(x)=\ln(1+x)$.

Objectives:
1. Use graphical methods to investigate the interval of convergence of a Taylor series.
2. Use graphical methods to find a bound for the error when a finite number of terms of a Taylor series are used to approximate a function over an interval.
3. Use the ratio test to find the interval of convergence.

Technology:
TI-82, TI-85, or TI-92 graphing calculator

Prerequisites:
1. Introduction to infinite series of constants.
2. Introduction to Taylor polynomials.
3. Introduction of the Ratio Test for infinite series of constants.

Overview: Given a sequence of numbers $\{a_n\}$, the nth partial sum s_n is the sum of the first n terms of $\{a_n\}$:
$$s_n = a_1 + a_2 + a_3 + \ldots + a_n$$

The sequence $\{s_n\}$ leads directly to an infinite sum $\sum_{n=1}^{+\infty} a_n$, also called an infinite series. The series $\sum_{n=1}^{+\infty} a_n$ converges if $\lim_{n \to +\infty} s_n$ exists, in which case the series is given the value L, where $L = \lim_{n \to +\infty} s_n$.

In this project we will extend the concept of an infinite series of constants to an infinite series of functions called a power series. For a power series $\sum_{n=1}^{+\infty} f_n(x)$ you will be asked to graphically explore values of x, say x=a, for which the infinite series of constants $\sum_{n=1}^{+\infty} f_n(a)$ converges.

Example 1: Consider the function $f(x) = \dfrac{1}{1+x}$ and the power series, $1 - x + x^2 - x^3 + x^4 \ldots (-1)^{n-1} x^{n-1} + \ldots = \sum_{n=1}^{+\infty} (-1)^{n-1} x^{n-1} = \sum_{n=1}^{+\infty} (-x)^{n-1}$. This series is obtained by long division of 1 by 1+x. To investigate values of x for which $\sum_{n=1}^{+\infty} (-x)^{n-1}$ converges, look at the sequence of functions $p_n(x)$ where
$$p_1(x) = 1$$
$$p_2(x) = 1 - x$$
$$p_3(x) = 1 - x + x^2, \text{ etc.}$$

If $\lim_{n \to +\infty} p_n(x)$ exists for a particular value of x then $\sum_{n=1}^{+\infty} (-x)^{n-1}$ converges for that value of x. The graphs of $P_5(x)$, $P_{10}(x)$ and $y = \dfrac{1}{1+x}$ are displayed in Figure 1 to give you an intuitive feeling that the sequence of polynomial functions $\{P_n(x)\}$ tends toward $\dfrac{1}{1+x}$ for $-1 < x < 1$.

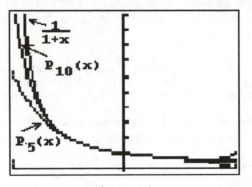

Figure 1

One might conjecture that $\sum_{n=1}^{+\infty} (-x)^{n-1}$ converges for any value of x where $-1 < x < 1$. For a value of x, say r, the series $\sum_{n=1}^{+\infty} (-r)^{n-1} = 1 - r + r^2 - r^3 + \ldots$ is a geometric series and converges if $|r| < 1$. It follows that the interval of convergence for the series $\sum_{n=1}^{+\infty} (-x)^{n-1}$ is $(-1, 1)$.

A series of the form $\sum_{n=0}^{+\infty} \frac{f^n(0) x^n}{n!}$, for a function $f(x)$ is called a Taylor series (Actually, this is a special type of Taylor series called a Maclaurin series). If x is in the interval of convergence, then this series converges to $f(x)$. That is

$$f(x) = f(0) + \frac{f'(0)}{1!}x + \frac{f''(0)}{2!}x^2 + \ldots + \frac{f^{(n)}(0)}{n!}x^n + \ldots$$

In this case, we can use the first n terms of this series, $P_n(x)$, to approximate $f(x)$. For x in the interval of convergence, the error in using $P_n(x)$ to approximate $f(x)$ is due to the fact that we are only using the first n terms, instead of an "infinite number" of terms. The size of the error, then, is the magnitude of those omitted terms.

$$f(x) = f(0) + \frac{f'(0)}{1!}x + \frac{f''(0)}{2!}x^2 + \ldots + \frac{f^{(n)}(0)}{n!}x^n + \frac{f^{(n+1)}(0)}{(n+1)!}x^{n+1} + \ldots$$

$$= P_n(x) + E_n(x)$$

$$P_n(x) = f(0) + \frac{f'(0)}{1}x + \frac{f''(0)}{2!}x^2 + \ldots + \frac{f^{(n)}(0)}{n!}x^n \text{ and the error is:}$$

$$E_n(x) = \frac{f^{(n+1)}(0)}{(n+1)!}x^{n+1} + \frac{f^{(n+2)}(0) x^{n+2}}{(n+2)!} + \ldots$$

In this project we will construct a Taylor series, graphically investigate its interval of convergence, and estimate a bound for the error when $p_n(x)$ is used to approximate $f(x)$.

Procedures:
I.
1. Find Taylor polynomials for $f(x) = \ln(x+1)$ about 0 of degrees 5 and 10.

2.   Find and record the Taylor series for $f(x)=\ln(x+1)$.

3.   **G1**   Print the graphs (see Appendix 4.1) of the Taylor polynomials of degrees 5 and 10 overlaid on the graph of $y=\ln(x+1)$. Label the graphs "P_5", "P_{10}" and "$\ln(x+1)$".

4.   The interval of convergence for a Taylor series is the set of all x-values for which $\lim_{n\to+\infty} P_n(x)=f(x)$. Hence, if x is in the interval of convergence, then $\lim_{n\to+\infty}(\text{error})=0$. Display the graphs of the error functions $E_5(x)=|\ln(x+1)-P_5(x)|$ and $E_{10}(x)=|\ln(x+1)-P_{10}(x)|$ on the same view screen. Based on these graphs, estimate the interval of convergence for this Taylor series (because this is only an estimate, do not concern yourselves with the endpoints of your interval).

II.
1.   Find Taylor polynomials for $f(x)=e^x$ about $x=0$ of degrees 5 and 10.

2.   Write the Taylor series for $f(x)=e^x$.

3.   **G2** Print the graphs of $P_5(x)$ and $P_{10}(x)$ overlaid on the graph of $y=e^x$. Label the graphs "P_5", "P_{10}", and e^x. Observe where each of the polynomials "look like" $f(x)=e^x$.

4.   Display the graphs of the error functions $E_5(x)=|e^x-P_5(x)|$ and $E_{10}(x)=|e^x-P_{10}(x)|$ on the same view screen. Based on these graphs, estimate the interval of convergence for this Taylor series.

III.
1.   Find Taylor polynomials for $f(x)=\cos x$ about 0 for degree 5 and 10.

2.   Write the Taylor series for $f(x)=\cos x$ about $x=0$.

3.   **G3** Print the graphs of $P_5(x)$ and $P_{10}(x)$ overlaid on the graph of $y=\cos x$. Label the graphs "P_5", "P_{10}" and $\cos x$. Observe where each of the polynomials "look like" $f(x)=\cos x$.

4.   Display the graphs of the error functions $E_5(x)=|\cos x-P_5(x)|$ and $E_{10}(x)=|\cos x-P_{10}(x)|$ on the same view screen. Based on these graphs, estimate the interval of convergence for this Taylor series.

IV.   Given that the Taylor series for $f(x)=\arctan x$ is $\sum_{n=0}^{+\infty} (-1)^n \frac{x^{2n+1}}{2n+1}$, use techniques similar to the ones in Procedure III. to estimate the interval of convergence.

V.   Graphical explorations of the interval of convergence of a series will not always lead to accurate conclusions. Analytic procedures are needed to provide precise results. The ratio test is usually used to find the interval of convergence for a power series such as $\sum_{n=0}^{+\infty} \frac{n}{3^n} x^n$. To use this test, one investigates the limit of the absolute value of the quotient of the (n+1)th term divided by the nth term.

$$\lim_{n \to +\infty} \left[\frac{n+1}{3^{n+1}} |x|^{n+1} \cdot \frac{3^n}{n|x|^n} \right] = \lim_{n \to +\infty} \left| \frac{n+1}{3^n} |x| \right|$$

$$= \frac{1}{3}|x|$$

The ratio test states that the series will converge for all values of x for which this limit is less than 1 and diverges for values of x for which the limit is greater than 1. Since the limit above is $\frac{1}{3}|x|$, the series will converge for all values x such that $\frac{1}{3}|x|<1$. Hence, the interval of convergence is $(-3,3)$. The ratio test does not provide information about the convergence at the end points x=-3 and x=3 of the interval and this project will not address this point.

1-4. For each series in Procedures I-IV, use the ratio test to determine the interval of convergence. Show your work and mark your answer clearly.

Checklist of calculator graph printouts to be handed in:
☐   **G1**   Print the graphs of the Taylor polynomials of degrees 5 and 10 overlaid on the graph of y=ln(x+1).
☐   **G2**   Print the graphs of $P_5(x)$ and $P_{10}(x)$ overlaid on the graph of $y=e^x$.
☐   **G3**   Print the graphs of $P_5(x)$ and $P_{10}(x)$ overlaid on the graph of y=cosx.

Objectives:
1.   Use a known power series for a function y=f(x) to generate a power series for a related function.
2.   Use a power series to approximate a definite integral and find a bound for the error.

Technology:
        TI-82, TI-85 or TI-92 graphing calculator

Prerequisites:
1.   Introduction to Taylor series.
2.   Differentiation and integration of trigonometric functions.
3.   Introduction to Integration and the Fundamental Theorem of Calculus.
4.   Introduction of the ratio test for power series.

Overview: Many times a definite integral contains an integrand whose antiderivative is unknown. Techniques exist for estimating these integrals such as calculating upper or lower sums, the trapezoid rule and the numerical integration capability of a calculator. The purpose of this project is to use power series to estimate such integrals and find an error bound for this estimate.

Procedures:
I.
1.   Find a Taylor series for cosx about x=0 and write your answer using summation notation.

2.   Use the ratio test to find the interval of convergence for this series.

3.   Use the Taylor series for cosx to write a series for $\cos(x^2)$ by substituting x^2 for x.

4.   Find the interval of convergence of the series for $\cos(x^2)$.

5.   Find and record the polynomial P(x) which is the first two terms of the power series for $\cos(x^2)$.   (P(x) should be a 4th degree polynomial).

6.   Find a bound B for the error when P(x) is used to approximate $\cos(x^2)$ on [0,.8] by using results about alternating series. Show your work.

II.

1.   Approximate the value of the definite integral $\int_0^{.8} \cos(x^2)\, dx$ by evaluating $\int_0^{.8} P(x)\, dx$.

2.   Use the bound B above to find a bound for the error when $\int_0^{.8} P(x)\, dx$ is used to approximate $\int_0^{.8} \cos(x^2)\, dx$. (The following information may be helpful: if $0 \le g(x) \le c$ for $a \le x \le b$ then

$$0 \le \int_a^b g(x)\, dx \le \int_a^b c\, dx.)$$

III.

1.   Write a Taylor series for $f(x) = \sin x$ about zero.

2.   Write a power series for $\dfrac{\sin x}{x}$ by using the Taylor series for $\sin x$.

3.   Write a polynomial $P(x)$ which is the first two terms of the series for $\dfrac{\sin x}{x}$.

4.   Use this polynomial to approximate $\displaystyle\int_{.5}^{1} \dfrac{\sin x}{x}\, dx$ by evaluating $\displaystyle\int_{.5}^{1} P(x)\, dx$.

5.   Find a bound for the error when $P(x)$ is used to approximate $\dfrac{\sin x}{x}$ on $[.5, 1]$ and use it to find a bound for the error when $\displaystyle\int_{.5}^{1} P(x)\, dx$ is used to approximate $\displaystyle\int_{.5}^{1} \dfrac{\sin x}{x}\, dx$.

IV.

When long division is used to divide 1 by $1+x^2$, the result is $\dfrac{1}{1+x^2} = 1 - x^2 + x^4 - x^6 + \ldots$ (or)

$$\frac{1}{1+x^2} = \sum_{n=0}^{+\infty} (-1)^n x^{2n}.$$

1.   Use the fact that $\dfrac{d(\arctan x)}{dx} = \dfrac{1}{1+x^2}$ to find a power series representation for arctan x.

2.   Write the polynomial $P(x)$ which is the first three terms of this series.

3.   Approximate the definite integral $\displaystyle\int_{0}^{.5} \arctan(x)\, dx$ by evaluating $\displaystyle\int_{0}^{.5} P(x)\, dx$.

4.   Find a bound for the error when $P(x)$ is used to approximate

arctanx on the interval [0,.5] and use it to find a bound for the error when $\int_0^{.5} P(x)\ dx$ is used to approximate $\int_0^{.5} \arctan(x)\ dx$.

Objectives:
1.   Use Newton's method to approximate the zeros of a function.
2.   Use Newton's method and a power series for e^x to approximate $\ln x$.

Technology:
      TI-82, TI-85 or TI-92 graphing calculator, program NEWTON.

Prerequisites:
1.   Find the equation of a line tangent to a function f at $(x_0, f(x_0))$.
2.   Find derivatives of the sine and exponential function.
3.   Introduction to Taylor series.

Overview: Appropriate mathematical computer software as well as advanced graphing calculator algorithms provide the math "power" to solve rather complicated problems.   We utilize technology to solve equations and evaluate functions with a few key strokes but we may not be aware of the mathematical "power" that is being used. The purpose of this project is to investigate how a calculator or computer can perform these operations.

Newton's Method will be used to approximate a zero, a, of a function f. To begin the process, an estimate for a is needed and we will designate it x_0, see Figure 1.

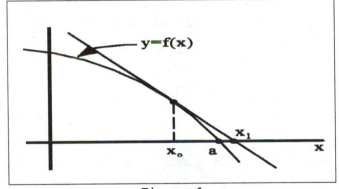

Figure 1

The equation of the line tangent to $y=f(x)$ at $(x_0, f(x_0))$ is $y - f(x_0) = f'(x_0)(x - x_0)$.   Let x_1 represent the solution to this equation when $y=0$.   Symbolically,

$$0 - f(x_0) = f'(x_0)(x_1 - x_0)$$

$$\text{so} \quad x_1 - x_0 = \frac{-f(x_0)}{f'(x_0)}$$

$$\text{or} \quad x_1 = x_0 - \frac{f(x_0)}{f'(x_0)}.$$

This intercept x_1 is the first calculated approximation for a by Newton's method.   Using this approximation for a, we can compute the second approximation

$$x_2 = x_1 - \frac{f(x_1)}{f'(x_1)} \quad \text{(see Figure 2)}.$$

A sequence of approximations $\{x_n\}$ is generated in this manner, according to the formula

$$x_n = x_{n-1} - \frac{f(x_{n-1})}{f'(x_{n-1})}.$$

This technique is called Newton's method. If the sequence $\{x_n\}$ converges to a, then each term of x_n is an approximation of a. If $m>n$, x_m is usually a better approximation than x_n. Figure 2 shows a graphical interpretation of

$$x_2 = x_1 - \frac{f(x_1)}{f'(x_1)}.$$

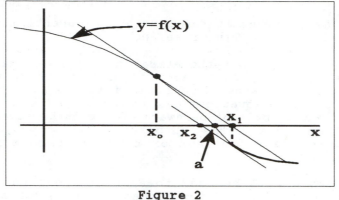

Figure 2

Procedures:
I.   The purpose of this procedure is to illustrate how to approximate a solution of the equation sinx=2x-1 by using a Taylor polynomial and Newton's method.
1.   Find a 9th degree Taylor polynomial, $P_9(x)$, for sin x about zero.

2.   Use a graph of y=sin x and y=2x-1 over the interval $\left[0, \frac{\pi}{2}\right]$ to make an eyeball guess for a solution, x_0, of the equation sinx=2x-1.

3.   Store $g(x)=P_9(x)-2x+1$ at y_1 and $g'(x)$ at y_2 on the y= menu (see Appendix 1.1). Execute the program NEWTON and enter your estimate x_0 from question 2 when prompted for a guess. Press **[ENTER]** and your guess x_0 and $g(x_0)$ will be displayed. At this time record these results in Table 1. Press **[ENTER]** again and record the value of x_1 and $g(x_1)$ in Table 1. Continue pressing the enter key and use the results to complete Table 1. (To exit the program at any time press [ON] and choose "2. Quit.")

n	x_n	$g(x_n)$
0		
1		
2		
3		
4		

Table 1

4. Discuss what role $g(x_n)$ plays in this problem. Describe the behavior of $g(x_n)$ as n increases and discuss the relevance of this behavior.

5. Use a built-in feature of your calculator to approximate the solution to sinx=2x-1 to eight decimal places (see Appendix 11.1). Does it appear that x_n, in the above table, converges to the solution? How many iterations of Newton's method were necessary to approximate the solution of sinx=2x-1 correct to five decimal places?

II. The purpose of this procedure is to construct a table of values for the natural logarithm function by evaluating polynomials. The Taylor series for e^x is easily constructed and converges for all real numbers. Because of its nice features, it will be used with Newton's method to approximate values of ln a for a=.1, .5, 1.1 and 2.

1. Find a Taylor series representation of e^x about zero and record $P_7(x)$, the first 7 terms of this series.

2. Given a number a where 0<a<e, we will approximate the solution to e^x=a by using Newton's method to approximate a solution of $P_7(x)$=a. For each value of a in Table 2 use part I as a guide to find x_0 thru x_4 and record these values in Table 2.

a	.1	.5	1.1	2
x_0				
x_1				
x_2				
x_3				
x_4				

Table 2

3.   For each value of a in Table 2, x_4 is an approximate solution of the equation $e^x=a$.

$$\text{Since } e^{x_4} \approx a$$

we can conclude that

$$\ln e^{x_4} \approx \ln a$$

$$\text{or } \ln a \approx x_4$$

Use this idea and the results of Table 2 to complete Table 3.

a	Approximations for ln a	Calculator values for ln a	Error
.1			
.5			
1.1			
2.0			

Table 3

4.   Suppose that you are programming in a language which is restricted to performing the four fundamental operations (addition, subtraction, multiplication, and division) on polynomials.   Discuss the merit of using this method to calculate values of ln x.   Be specific.

Objectives:
    Given a set of points which are not colinear, find the equation of the line which "best" fits the given points.

Technology:
    TI-82, TI-85, or TI-92 graphing calculating, program RMSE

Prerequisites:
    A study of linear functions

Overview: One widely used application of mathematics is to "fit" a curve to a set of data when an exact fit doesn't exist. The purpose of curve fitting is to determine if there is a relationship between two or more variables even though the relationship may not be exact. The relationship may be linear, curvilinear, or there may be no relationship at all. One might conjecture that the number of new housing starts in South Florida depends on the rate of interest. The independent variable x is the interest rate and the dependent variable y is the number of new housing starts. After fitting a curve to a given set of data, the equation of the curve is usually used to estimate the value of the dependent variable y for a given value of x.

A graphing calculator will be used to plot a scatter diagram. The scatter diagram will be used to get an idea about the type of relationship which exists between the variables and to find the curve which best fits the data.

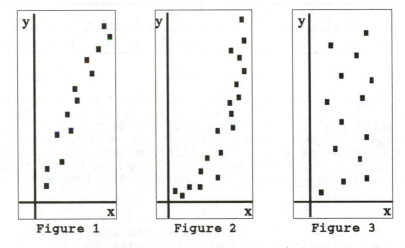

Figure 1   Figure 2   Figure 3

The scatter plot in Figure 1 indicates a linear relationship between x and y, the one in Figure 2 represents a curvilinear relationship, and no relationship exists between the variables in Figure 3. A scatter plot is used to get a general impression about an appropriate model. It is rare to find a perfect fit to describe a relationship (In a perfect fit, all points lie on the curve). In this project we will consider only linear relationships.

Example 1: One method for finding a linear model for a given data set is to plot a scatter diagram, guess the equation of the line, graph that equation, and based on the graph of the line and the given points, make an adjustment to the equation to obtain a "better fit". When using this "eyeball" technique it would be helpful to have a precise criterion for choosing the "best" model

when several are being considered.1 To develop a test criterion
we will consider the data in Table 1 and the line whose equation
is y=2.4x-2.

x	1	2	3	4	5
y	1.4	4.0	4.3	7.6	9.1

Table 1

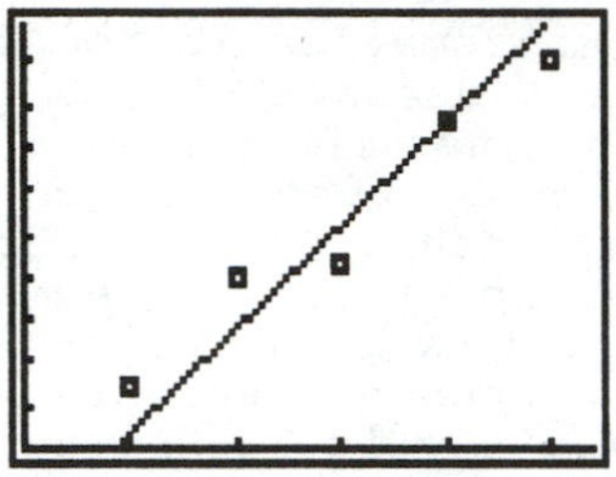

Figure 4

Figure 4 shows a scatter plot of these
points and the graph of the equation
y=2.4x-2.   To use this linear model,
one uses the equation to "find" y for a
given value of x. For example when x=2
the calculated y-value, which we will
call $\hat{y}$, has a value of 2.8.   The error
between the calculated value $\hat{y}$ and the
exact value of y from Table 1 is
$\hat{y}$-y=2.8-4=-1.2 .

x	1	2	3	4	5
y	1.4	4.0	4.3	7.6	9.1
$\hat{y}$	0.4	2.8	5.2	7.6	10.0
$\hat{y}$-y	-1.0	-1.2	0.9	0	0.9
$(\hat{y}$-y$)^2$	1.0	1.44	0.81	0	0.81

Table 2

The coordinates of the five points from Table 1 are listed in the
first two rows of Table 2.   The values of $\hat{y}$ = 2.4x - 2 are listed
in the third row.   The fourth row is the difference, $\hat{y}$-y, which
represents the error when using $\hat{y}$ to approximate y for a
particular value of x, and the fifth row contains the square of
the errors.   One way to measure how well a line fits a given data
set is to calculate the root mean square error (rms error), which

$$\text{is:} \quad \sqrt{\frac{(-1)^2+(-1.2)^2+(.9)^2+0^2+(.9)^2}{5}} = \sqrt{\frac{4.06}{5}} = .90 .$$

The rms error is a measure of how well the model fits the actual
data.   The smaller the rms error the better the fit.

If one were to use the model $\hat{y}$ = 2x-1 for the same data set, the
root mean square error would be .63.   Therefore, $\hat{y}$ = 2X-1 is
considered a better fit than $\hat{y}$ = 2.4x - 2 since the rms error for
$\hat{y}$ = 2x-1 is smaller.

Procedures:
I.   For each of the given data sets a-d below, each member of
your team is to:
1.   Store the x-values in a list L_1 and the y-values in a list
L_2.   See Appendices for details about list storage.
2.   Graph a scatter plot of the data set.
3.   Guess a linear model, store it at y_1 on the y= menu, and
overlay the graph of this line on the scatter plot.

4.   Make adjustments to your linear model, regraph, and compare this graph with the scatter plot.   Continue this adjustment process until you have a "good" fit. **G1-G4** Print the graph of your best fit overlaid on a scatter plot of the data.
5.   Calculate the rms error by running the program RMSE. This program will perform the computation's outlines in example 1 and display the rms error for your model.   To run the program, your equation must be stored at y_1, x-values in list L_1, and y-values in list L_2.
6.   Record each person's model and their corresponding rms error.   Determine which member of your team has the best fitting linear model.   How did you decide?

a. $\{(1,-3.2),(2,-3.3),(4,-3.6),(5,-2.8),(-2,-4.1)\}$
b. $\{(1,-1.4),(2,-3.5),(5,-8.1),(-2,5),(-5,6)\}$
c. $\{(1,1),(4,7),(2,3),(-2,-5),(-5,-11)\}$
d. $\{(-1,1),(2,0),(4,-5),(-2,12),(3,-8)\}$

II.   We will now use the linear regression mode on your calculator to find a best fit for these data sets.   For each of the data sets (a-d) in Part I, each member of your team is to perform linear regression on the data set (see Appendix 7.2 for details on linear regression) and record the regression equation, correlation coefficients and the rms error.   (To calculate the rms error, put the regression equation in y_1, x-values in list L_1, y-values in list L_2, and run the RMSE program.)

III.   Use the results from part II to describe any relationship between how well the model fits the data and the magnitudes of the rms error and correlation coefficient.

Checklist of calculator graph printouts to be handed in:
☐   **G1-G4** Print the graph of your best fit overlaid on a scatter plot of the data.

PROJECT 49 - Life Expectancy 1

Objectives:
1. Investigate the life expectancy trend for males and females in the United States.
2. Fit a line to each set of data where time (in years) is the independent variable and life expectancy, or average life span, is the dependent variable.
3. Answer questions and make predictions based on the linear models.

Technology:
TI-82, TI-85, or TI-92 graphing calculator.

Prerequisites:
A study of linear functions.

Procedures:
1. Enter the data for "year born" from Table 1 (taken from The World Almanac 1993 by Funk and Wagnalls Corp.) in list L1 on your calculator (see Appendix 6.1) and "life expectancy/male" in L2. Use linear regression (see Appendix 7.2) to find and record a linear model for male life expectancy, where the independent variable is the year born.

2. Enter the data for "life expectancy/female" in list L3 on your calculator and perform linear regression on L1 and L3. Record your linear model for female life expectancy.

3. **G1** Print the graphs (see Appendix 4.1) of both equations overlaid on scatter plots of the data on the same screen (see Appendix 1.6). Label the graphs "male" and "female".

4. What is the trend for life expectancy of females in the U.S.? For males?

5. Is the rate of increase or decrease in the life expectancy greater for males or females?

6. Estimate the life expectancy for a male and for a female born in 1991.

7. If this trend continues, in what year will a male and female have to be born in order to have a life expectancy of 80 years?

8. If this trend continues, in what year of birth will the life expectancy of a female and male be the same?

Year born	1976	1977	1978	1979	1980	1981	1982	1983	1984	1985	1986
life exp/ Male	69.1	69.5	69.6	70.0	70.0	70.4	70.9	71.0	71.2	71.2	71.3
life exp/ Fem.	76.8	77.2	77.3	77.8	77.4	77.8	78.1	78.1	78.2	78.2	78.3

Table 1

Checklist of calculator graphs to be handed in:

☐   **G1**   Graph both male and female life expectancy equations
                on the same screen.

Objectives:
1. Use regression to fit a line to the data.
2. Use the line to make predictions.

Technology:
    TI-82, TI-85, or TI-92 graphing calculator.

Prerequisites:
    An introduction to linear and polynomial regression.

Overview: The record for a one mile run set in 1913 was 4:14.4
and the record dropped below four minutes for the first time in
1954. Once the four minute barrier was removed, a mile run of
less than four minutes was no longer a rare event. The records
for a mile run and the years of each record are in Table 1[1]. The
purpose of this project is to fit a line to this data and use
this line to look for trends and make predictions.

Year	Mile record	Year	Mile record
1913	4:14.4	1954	3:58.0
1915	4:12.6	1957	3:57.2
1923	4:10.4	1958	3:58.5
1931	4:09.2	1962	3:54.4
1933	4:07.6	1964	3:54.1
1934	4:06.8	1965	3:53.6
1937	4:06.4	1966	3:51.3
1942	4:04.6	1967	3:51.1
1943	4:02.6	1975	3:49.4
1944	4:01.6	1979	3:49.0
1945	4:01.4	1980	3:48.8

Table 1

Procedures: In order to look for a relationship between years and
record times, change the times to seconds and carry your
calculations out to one tenth of a second.

I.
1.   With years as the independent variable and time of the
record in seconds as the dependent variable, use the linear
regression feature on your calculator to fit a linear model to
the data (see Appendix 7.2). Record your model.

2.   What is the slope of your regression line?

---

[1]From <u>The Complete Book of Track and Field</u> by Tom McNab, Exeter
Book, N.Y.

3.   Give an interpretation of the slope for this particular application.

4.   Do you think that your model will accurately predict the world record time for 1993? Why or why not?

5.   Use your model to predict the world record time for the mile run in 1993.

6.   The actual world record time in 1993 was 3:44.4[2]. Determine the size of the error in your prediction using the formula |error| = |actual time - predicted|.

II.   The model you derived in Procedure I represents the overall trend in the data from 1913 to 1980. Sometimes, however, the trend changes. In such cases, a model using all of the data points may not be the best model for prediction. Figure 1 shows an example of a data set whose trend changes at t=t*.

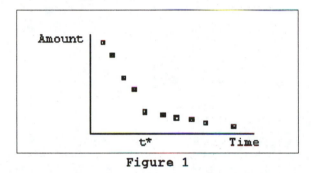

Figure 1

In this example, we would not want to use all of the data points to predict future amounts. Because the trend of the data changes at t=t*, we would want to use only those points where t≥t* to make predictions.

1.   Look at the scatter plot for the mile run data. If you notice a subtle change in the trend for the data at some point t*, then use only those points where t≥t* to calculate a regression equation. Record your choice of t* and the equation for your model.

2.   Using this model, do you get a better prediction for 1993?

3.   What is the slope of this regression line?

4.   How does it compare with the slope of the model in Procedure I which included all years, 1913 to 1980?

---

[2]From <u>The World Almanac 1995</u> by Funk & Wagnalls Corp.

5.   Give an interpretation of the difference in slopes for this application.

6.   Do you think that the world record for the mile run will ever reach 3:30.0?

7.   Use one of your models to predict when this will happen.

8.   Why did you use this model to make your prediction?

Objectives:
1.   Experimentally discover Hooke's Law.
2.   Determine a method for describing springs.

Technology:
        TI-82, TI-85, or TI-92 graphing calculator

Equipment:
1.   3 springs whose lengths are between 30cm and 80cm.
2.   Five masses between 100g and 600g.
3.   A meter stick.

Overview: When a mass is attached to the end of a hanging spring, a force is exerted on the spring and it is displaced s units from its initial position (see Figure 1).

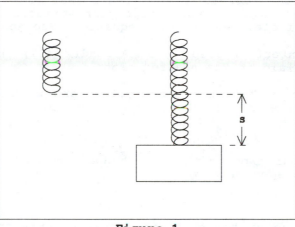

Figure 1

The purpose of this experiment is to find a relationship between the displacement of a spring and the resulting spring force.

Procedures:
I.
1.   Attach a spring to the ceiling or a rod so that the end of the spring is at least 1.5 meters from the floor. Measure and record the distance from the end of the spring to the floor in centimeters.

2.   Attach a known mass to the end of the spring and measure the distance from the end of the spring to the floor. Repeat this procedure with 4 other masses and fill in the first two columns of Table 1.

3.   Complete Table 1 (Be careful with the units). To find the force (in the third column) use the formula F=9.81m, where mass m is in kilograms and force F is in Newtons (the constant $9.81m/sec^2$ is the gravity constant).

4.   Store the numbers in columns 3 and 4 of Table 1 in lists L1 and L2 on your calculator (see Appendix 7.1).

5.   Graph a scatter plot of displacement and force where the independent variable x is displacement and the dependent variable y is force (see Appendix 1.5).

	mass in kg	distance from the floor in meters	force in Newtons	displacement s in m (see Figure 1)
1				
2				
3				
4				
5				

Table 1

6. Use linear regression to fit a linear model to this data with displacement being the independent variable and force the dependent variable. Record your equation (See Appendix 7.2).

7. **G1** Printout a graph (see Appendix 4.1) of the linear equation overlaid on the scatter plot.

8. Look up Hooke's Law (of physics) and write the formula. Explain what the letters in Hooke's Law mean.

9. Does your experiment support Hooke's Law? Explain your answer and be specific.

10. The value of the constant, usually represented by k, in Hooke's Law is called the spring constant. What is the spring constant for your spring?

11. Use the model for your spring to fill in Table 2.

mass in kilograms	displacement in meters
.3	
.4	
	.25
	4.0
3.0	
	.6

Table 2

In each case, is the given input (kilograms or meters) appropriate for your spring? Explain.

II.
1.   Repeat steps 2 in Part I for two springs which are different from the spring used in Part I and find the spring constant for each one.

2.   Pull each spring and rank them from one to three, where one is the "weakest" and three is the "strongest" spring. What is the relationship between your strength ratings and the spring constants?

3.   A person is going to install coil springs on a truck used for heavy hauling and has a choice of three types of springs. They want the strongest spring possible and the spring constants are written on the box. How should this person make the choice?

4.   Your manual garage door really flies up and is hard to close. You notice that this is caused by the two springs which are attached to the door. Describe what you would do if you decide to purchase new springs so that the garage door will be easier to pull down. Suppose that you purchase garage door springs by their spring constants.

Checklist of calculator graph printouts to be handed in:
☐   **G1**   Printout a graph of the linear equation overlaid on the scatter plot.

Objectives:
1.   Use the natural logarithm function to transform a data set
     which appears to be exponential. Determine if the
     transformed data looks linear; if so, the data is said to
     be linearized.
2.   Fit a linearized data set with a linear model and use the
     linear model to find an exponential model for the original
     data set.
3.   Use the exponential regression algorithm on your calculator
     to find an exponential model directly.

Technology:
     TI-82, TI-85, or TI-92 graphing calculator

Prerequisites:
     Introduction to the natural logarithm function and its
     inverse function, $y=e^x$.

Overview: Consider the data in Table 1, which was obtained by
measuring the number of bacteria in a culture for 7 consecutive
days.

Time elapsed in days	Number of bacteria
0	600
1	670
2	748
3	834
4	931
5	1039
6	1161

Table 1

A scatter plot of this data in Figure 1 indicates that it is
reasonable to fit a linear model to this data.

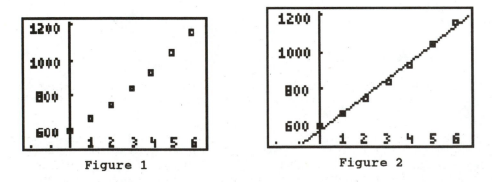

Figure 1   Figure 2

A scatter plot and the regression line are shown in Figure 2.
The fit looks good and a correlation coefficient of .995 provides
further evidence that this line is a good fit. However, one week
later on the 14th day, a count of 2799 was found.

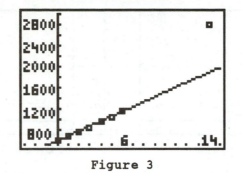

Figure 3

As you can see in Figure 3, the linear model generated from the first seven days is a very poor predictor of the number of bacteria on day 14.   The linear model generated by the first seven days predicted a count of 1878 for day 14, when the actual count was 2799.   There are two lessons to be learned from this experience.   The first is that one has to be careful about using a model to make predictions when there is a large jump in the value of the independent variable.   In this case, the model was based on the first seven days and our prediction was for the 14th day.   Secondly, a linear model is usually not appropriate for a growth application.

(Skip this next paragraph if you haven't studied derivatives and integrals.)

For the linear model, y = ax + b, the rate of change in y with respect to x is the constant a, which is the slope of the line. In many growth applications, the rate of change of a population with respect to time is not constant because as the population increases, the growth rate usually increases also.   The growth rate usually depends on the size of the population and sometimes the growth rate is assumed to be proportional to the size of the population.   Let p = f(t), where t is time and p is the number of individuals in the population at time t.   If we assume that the growth rate is proportional to the size of the population, then

$$\frac{dp}{dt}=kp$$

or $$\frac{dp}{p}=kdt$$

integrating, $\ln p=kt+c$
therefore $P=e^{kt+c}$
or $P=e^c e^{kt}$
or $P=P_0 e^{kt}$
Where P_0 is the initial population.

This model is based on the assumption that the rate of change in the population is proportional to the size of the population at time t.

The "eyeball" method is not very reliable for deciding whether a set of data is exponential.   The graphs in Figures 4 and 5 appear to be members of the same family.   However, the function in Figure 4 is the exponential function $y = e^x$, whereas the one in Figure 5 is the cubic function $y= 2x^3+1$.   A more reliable way of deciding whether our data set is exponential is to try to "linearize" the data.

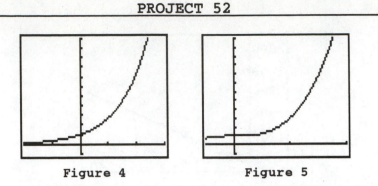

Figure 4   **Figure 5**

Given a set of data, we will use a logarithm function to transform the data. If the transformed data "looks" linear, we will say that the transformation linearized the data and we will fit it with a regression line. Consider the exponential model:

$$y=ae^{kx}$$

Exponentiating, $\ln y=\ln(ae^{kx})$

so, $\ln y=\ln a + \ln e^{kx}$

or $\ln y=\ln a + kx$

Therefore if you start with an exponential model, then $\ln y$ is a linear function of x because k and $\ln a$ are constants. Table 2 contains the bacteria count, y, along with $\ln y$, for the data from Table 1.

x	Number of bacteria,y	ln y
0	600	6.3969
1	670	6.5073
2	748	6.6174
3	834	6.7262
4	931	6.8363
5	1039	6.9460
6	1161	7.0570

Table 2

To determine if an exponential model is appropriate for the original data, we will determine if a linear model is appropriate for the transformed data $(x, \ln y)$. A scatter plot of the transformed data and a graph of the regression line are shown in Figure 6.

The transformed data looks linear and the correlation coefficient is .999998, so it appears as though the original data was exponential. Since $f(x) = e^x$ is the inverse of $g(x) = \ln x$, $f(x)$ will be used to transform the linear model to an exponential model of the form $y = p_0 e^{kt}$. From the linear regression we get a slope of .10988 and an intercept of 6.39707 so

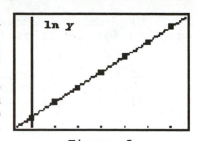

Figure 6

$$\ln y = .10988x + 6.39709 \qquad (1)$$
$$e^{\ln(y)} = e^{.10988x + 6.39709} \qquad (2)$$
$$y = (e^{.10988x})(e^{6.39709}) \qquad (3)$$
$$y = 600.1\, e^{.10988x} \qquad (4)$$

Note that $p_0 = 600.1$ is close to the initial count of 600. When we used a linear model of the original data the predicted count for day 14 was 1878, while the actual count was 2799. The count obtained from the exponential model, $y = 600.1 e^{.10988x}$, for day 14 is 2794. This model appears to be more suitable for bacteria growth, within the time frame of this experiment, than the linear model.

Procedures:

I. You invest some money which is compounded continuously at a fixed annual rate. The balance at the end of each of the first six years is: $5443.60, $5926.50, $6452.30, $7024.70, $7648.00, and $8326.50, respectively.

1. Enter the time (in years) in List L1 on your calculator and the balance at the end of each of these years in List L2 (see Appendix 7.1). Linearize this data by taking the natural logarithm of the balances (see Appendix 6.2). Use your calculator to fit a line to this data (see Appendix 7.1).

   a. Record the regression equation and the corresponding correlation coefficient.

   b. Use equations 1-4 in the previous example as a guide to find and record an exponential model of the form $y=ae^{kx}$ for the original data.

   c. **G1** Printout a graph (see Appendix 4.1) of this exponential model overlaid on a scatter plot of the original data (see Appendix 1.6).

   d. What is your initial investment?

   e. What is the balance at the end of 12 years?

II. Table 3 below provides information about inflation and the buying power of $100 based on 1982 dollars (taken from the Statistical Abstracts of the United States, 1994). Linearize the data and use linear regression to fit this data with a line. Exponentiate to find a model of the form $y=ae^{kx}$ where x is the number of years past 1980 and y is the buying power of $100 based on 1980 dollars (i.e. x=0 corresponds to 1980).

1. Record the regression equation for the linearized data.

2. Use this equation to find and record the exponential equation for the original data.

3. **G2** Printout a graph of the exponential model overlaid on a scatter plot of the original data.

4. If this trend continues when will the buying power be only $50?

5. If this trend continues, what will the buying power be in the year 2010?

Year	Buying power
1983	$98.40
1985	$95.55
1987	$94.90
1989	$88.20
1991	$82.20

Table 3

III. In population analysis, for example, the factors that affect the size of the population will change with time, so that no single model can provide accurate results over large spans of time. In this procedure we will address this issue by constructing models on smaller time intervals.

Table 4 contains world population estimates in millions from 1650 to 1991 (taken from <u>The World Almanac</u>, 1993).

Year	Population in millions
1650	550
1750	725
1850	1175
1900	1600
1950	2564
1980	4478
1991	5423

Table 4

Use the restrictions in problems 1-3 and the data in Table 4 to find exponential models of the form $y = ae^{kx}$ where x is the number of years past 1650 and y is the population of the world in millions (i.e. x=0 corresponds to the year 1650).

1. Suppose a statistician in 1905 had used the data from 1650-

1900 to find an exponential growth model.
      a.   What model would he have found?

      b.   What would his population predictions have been for 1950, 1980 and 1991?

      c.   Calculate the error for each of these predictions based on the actual population in Table 4.

2.   Use the data from 1650 through 1980 to derive and record an exponential model and use it to predict the 1991 population.

3.   **a.**   Find and record an exponential model by using the data from 1850 through 1980.

      b.   Use this model to predict the population in 1991.

4.   **G3** Print out a graph of the above three exponential models overlaid on the scatter plot of the data in Table 4. Label the graphs "#1", "#2", and "#3" corresponding to parts 1-3 above.

5.

      a.   Use the model that you have the most confidence in for predicting the population for the year 2100.

      b.   Do you think this estimate is too high or too low? Defend your answer.

      c.   How many times larger than the population in 1991 is your population prediction for 2100?

      d.   Assume that we would have to multiply the food supply for 1991 by this amount to feed the population in 2100: Do you think this is possible? If not, what do you think will happen?

IV.
Which of the models in this project would you place the most confidence in for making predictions.
a.   The world population model
b.   The interest model
c.   The buying power model

Rank them in the order of highest confidence to lowest confidence and defend your answer.

V.
1.   Your calculator has a built in exponential regression feature for fitting data with an exponential model. Use this feature to fit an exponential model to the data in Table 3 and record your result (see Appendix 7.7).

2.   Compare this exponential equation with the one in Procedure II and algebraically show that the two are equivalent (i.e. it is not sufficient to say that the graphs look the same).

Checklist of calculator graph printouts to be handed in:

☐   **G1**   Printout a graph of this exponential model overlaid on a scatter plot of the original data (interest model).

☐   **G2**   Printout a graph of the exponential model overlaid on a scatter plot of the original data (buying power model).

☐   **G3**   Printout a graph of the above three exponential models overlaid on the scatter plot of the data in Table 4 (world population model).

Objectives:
1.   Linearize a data set using logarithms.
2.   After fitting a linearized data set using linear regression, find the power function which describes the original data set.
3.   Determine the relationship between the LPS and the concentration of bacteria in a lake or ocean.
4.   Determine the reaction order of a chemical reaction.
5.   Determine the relationship between the drag coefficient and Reynolds number.

Technology:
   TI-82, TI-85, or TI-92 graphing calculator

Prerequisites:
   Introduction to the natural logarithm function and its inverse function, the exponential function $y=e^x$

Overview: The power function $y=kx^c$ can be transformed to a linear function by taking the natural logarithm of both sides:
$$y=kx^c$$
$$\ln y=\ln(kx^c)$$
$$\ln y=\ln k + \ln x^c$$
$$\ln y=\ln k + c \ln x$$
Now $\ln y$ is a linear function of $\ln x$, since c and $\ln k$ are constants.

Procedures:
1.   Chemists use power regression to determine the reaction order of certain chemical reactions. Table 1 contains data from a series of experiments with different concentrations of iodide when iodide ions are oxidized by peroxide sulfate.[1]

iodide concentration, [I⁻]	rate, v
0.037	.00071
0.057	.00152
0.074	.00143
0.095	.00173
0.110	.00201

Table 1

The order of this reaction corresponds to the exponent, c, in the formula
$$v=k[I^-]^c$$
Here the "-" sign refers to ions and the brackets refer to concentration, so the expression $[I^-]$ refers to the concentration of iodide ions.   The rate, v, is the rate at which the concentration of peroxide sulfate is changing with respect to time.

---

[1]"Microemulsions as a New Working Medium in Physical Chemistry" by Julia Casado and Maria Luisa Maya in <u>Chemical Education Journal</u> May, 1994.

This formula can be linearized by taking the natural logarithm of both sides:

$$v = k[I^-]^c$$
$$\ln v = \ln (k[I^-]^c)$$
$$\ln v = \ln k + \ln[I^-]^c$$
$$(1) \quad \ln v = \ln k + c \ln [I^-]$$

Now $\ln v$ is a linear function of $\ln [I^-]$, since $\ln k$ and c are both constants throughout the experiment.

Take the natural logarithm of the numbers in each column of Table 1 (see Appendix 6.4). Use linear regression on this transformed data to find the values of c and $\ln k$ in equation (1) (see Appendix 7.2).

a. Record your linear equation in the form of equation (1), substituting your values for $\ln k$ and c.

b. Based upon the values of your parameters, write v as a function of $[I^-]$.

c. What is the order of the reaction, c?

d. What is the rate constant, k?

G1 Print a graph of your function overlaid on a scatterplot of the data in Table 1 (see Appendix 4.1).

2. Biologists have hypothesized that a power function relation exists between the concentration of LPS and the concentration of bacteria in lakes and oceans. If the parameters for this relationship can be accurately determined, then we can estimate the number of microorganisms in lake and ocean waters by checking the LPS level. The data in Table 2[2] has been collected to determine the parameters k and c for the relationship $(Bacteria/ml) = k(LPS/ml)^c$. Linearize the data in Table 2 and use linear regression to determine k and c. Use these parameters to write bacteria/ml as a function of LPS/ml.

a. Record your formula.

G2 Print a graph of your function overlaid on a scatterplot of the data in Table 2.

b. Suppose that the LPS level for a particular location in the Florida Keys was measured at 16,000 LPS/ml. Use your function to estimate the number of bacteria per ml for this location.

---

[2]Data from <u>Microbial Ecology</u> by Atlas.

LPS/ml	Bacteria/ml
67	16318
90	18034
200	66171
191	98716
493	179872
992	442413
2697	1202604
8103	3269017
8103	5956538
12088	6920510

Table 2

3.   In engineering, the study of fluid mechanics is essential for analyzing the flow of air and other gases through automobile and airplane engines, the flow of air around an airplane's wings, and the lift that is generated from this flow. The study of such flows requires a knowledge of the relationship between the drag coefficient, c_D, and the Reynolds number Re_L. For turbulent fluid flow across a smooth flat plate, calculus can be used in conjunction with engineering theory to derive the relationship

$$c_D = \frac{0.072}{Re_L^{1/5}}.$$

As Fox and McDonald report in their popular book *Introduction to Fluid Mechanics*, a better fit can be obtained with experimental data. Use the data in Table 3[3] to obtain such a "better fit". Record your results.

---

[3]Data from <u>Introduction to Fluid Mechanics</u> by Fox and McDonald

Re$_L$	C$_D$
290000	.00598
450000	.00548
600000	.00517
820000	.00486
1200000	.00450
3800000	.00358
7400000	.00312
8900000	.00302

Table 3

G3   Print a graph of your function overlaid on a scatter plot of the data in Table 3.

Note: Engineers use this relationship to determine the drag coefficient for a particular Reynolds number.   Thus, curve-fitting is being used here to interpolate between experimental data points rather than to make predictions.

4.   Your calculator has a built-in power regression feature. Use it to fit a power function to the data in Table 3.   Record your result (see Appendix 7.8).

Since the built-in power regression feature on your calculator follows the same algorithm as you did, by linearizing the data and then applying linear regression, your result should be identical to the result from procedure 3.

Checklist of calculator graph printouts to be handed in:
☐   **G1**   Print a graph of your function overlaid on a scatter plot of the data in Table 1 (Chemistry data).
☐   **G2**   Print a graph of your function overlaid on a scatter plot of the data in Table 2 (Biology data).
☐   **G3**   Print a graph of your function overlaid on a scatter plot of the data in Table 3 (Engineering data).

Objectives:
1. Develop a growth model with an assumption that there is a bound for the population size.
2. Use logarithms to linearize data and determine the parameters by linear regression.

Technology:
TI-82, TI-85, or TI-92 graphing calculator

Prerequisites:
1. Basic knowledge of integration
2. Basic knowledge of points of inflection
3. Basic knowledge of the natural logarithm function

Overview: The data in Table 1 is world population in millions of people from 1900 to 1991 (taken from <u>The World Almanac</u> 1993 by Funk and Wagnalls Corp.).

Year, x	Population in millions, p
1900	1,600
1950	2,564
1980	4,478
1991	5,432

Table 1

An exponential model for this population data is $p = 1506(1.0135)^x$. According to this model the population in 2100 is 21,965 million, which is four times the population in 1991. This does not seem to be reasonable, as a population of this size would certainly strain our water supply, and without major changes in food production, the food supply would not sustain this many people. It is not reasonable to expect the number of people in the world to grow exponentially without bound.

The underlying assumption for the exponential model is that the rate of growth is proportional to the size of the population. The exponential model, $p(x) = p_o \, e^{kx}$, is derived by solving the differential equation $\frac{dp}{dx} = kp$, with $p(0) = p_o$. A model of population growth developed by Verhulst assumes that there is an upper bound L for the population, and as the population approaches L, the rate of growth will decrease. His model was derived from the differential equation $\frac{dp}{dx} = kp\frac{L-p}{L}$. When p is small $\frac{L-p}{L}$ is close to 1 and the graph looks like the graph of $p = p_o e^{kx}$. So the graph appears exponential first, with k corresponding to the growth coefficient in the exponential model $p = p_o e^{kx}$. When p approaches L, $\frac{L-p}{L}$ approaches 0, so the expected growth rate approaches 0, and the graph has a horizontal asymptote p=L.

To investigate a graphical solution of $\frac{dp}{dx} = kp\frac{L-p}{L}$, we will first approximate L. Suppose the population begins to level off in 1991. For reasons that will be explained later, let L be twice the population in 1991 or 10,846 million. A graphical solution of $\frac{dp}{dx}=.0134p\frac{10846-p}{10846}$ with $p_o=1{,}600$ million is displayed in Figure 1.

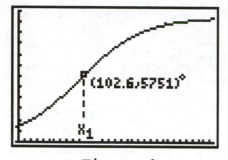

Figure 1

The part of the curve over the interval $(0,x_1)$ looks exponential and the growth rate is greatest at x_1. At this point the growth rate changes from increasing to decreasing. Therefore a point of inflection occurs at x_1. The model $p=\dfrac{L}{1+e^{ax+b}}$ is one form for the family of functions with the characteristics of the graph in Figure 1. This model, called a logistic growth model, is more difficult to analyze than a linear model $y=ax+b$ or an exponential model $y=ae^{kx}$ because it contains three parameters: a, b, and L. We will devise a method to estimate L and then linearize the data and use linear regression to estimate a and b. Below is a derivation of the relationship between the height of the inflection point and the parameter L. We will then use this relationship to estimate L when fitting a curve.

$$p=\frac{L}{1+e^{ax+b}}$$

$$p'=\frac{(1+e^{ax+b})\,0-L(ae^{ax+b})}{(1+e^{ax+b})^2}=\frac{-Lae^{ax+b}}{(1+e^{ax+b})^2}$$

$$p''=\frac{(1+e^{ax+b})^2\,(-La^2e^{ax+b})+Lae^{ax+b}.2(1+e^{ax+b})\,(ae^{ax+b})}{(1+e^{ax+b})^4}$$

$$=\frac{(1+e^{ax+b})\,(La^2e^{ax+b})\,[-(1+e^{ax+b})+2e^{ax+b}]}{(1+e^{ax+b})^4}$$

$$=\frac{(La^2e^{ax+b})\,[e^{ax+b}-1]}{(1+e^{ax+b})^3}$$

Looking for the inflection point, p'' is never undefined, $L\neq0$ and $a\neq0$. Therefore:

$$p''=0 \quad\leftrightarrow\quad e^{ax+b}-1=0$$
$$\leftrightarrow\quad e^{ax+b}=1$$
$$\leftrightarrow\quad \ln(e^{ax+b})=0$$

$$\hookrightarrow \quad ax+b = 0$$
$$\hookrightarrow \quad x = -\frac{b}{a}$$

Since $\frac{-b}{a}$ is the only value of x for a possible point of inflection, and the previous discussion indicated that the growth rate changes from increasing to decreasing, one can conclude that a point of inflection occurs at $x = \frac{-b}{a}$.

$$p\left(\frac{-b}{a}\right) = \frac{L}{1+e^{a\left(\frac{-b}{a}\right)+b}} = \frac{L}{1+1} = \frac{L}{2}.$$

We can conclude that a model of the form $p = \frac{L}{1+e^{ax+b}}$ has a point of inflection at (-b/a , L/2). This is the point where the growth rate is greatest and also the point at which the rate of growth begins to slow down. If we can estimate the population at the point of inflection, we can use it to obtain an estimate for the parameter L. The point of inflection (-b/a , L/2) provides a reasonable estimate for L, but attempting to estimate a and b from the point of inflection will give inconsistent results. This is because small variations in a and b will drastically affect p, while p is not as sensitive to small changes in L. Read Example 1 carefully since you will use this technique for determining a and b.

Example 1: Consider again the world population data in Table 1 and the exponential model $p=1506e^{.0134x}$ for this data. Assume an authority on world population trends has predicted that the world population growth rate will begin to slow down in the year 2000. We will use this information to fit a logistic growth model to the data. An estimate of the world population in the year 2000 from the exponential model is 5,751.5 million. Since $p = \frac{L}{2}$ at the point of inflection it follows that: 5,751.5 million $= \frac{L}{2}$ therefore L=11,503 million. The logistic model $p = \frac{11503}{1+e^{ax+b}}$ will be complete after we estimate the parameters a and b. To linearize the data, we will transform the equation:

$$p = \frac{11503}{1+e^{ax+b}}$$
$$p+pe^{ax+b} = 11503$$
$$e^{ax+b} = \frac{11503-p}{p}$$
$$ax+b = \ln\left(\frac{11503-p}{p}\right)$$

Let y=ln[(11503-p)/p]; then y=ax+b is a linear equation and we can use linear regression to determine a and b.

Table 2 contains values for ln[(11503-p)/p] (see Appendix 6.2).

Year	x	p (in millions)	y=ln[(11503-p)/p]

1900	0	1600	1.8228
1950	50	2564	1.2489
1980	80	4478	0.4503
1991	91	5423	0.11436
2000	100	5751.5 (estimate)	0

Table 2

Linear regression (see Appendix 7.2) on y vs. x produces the equation $y=-0.01895x+1.94411$. Hence a=-0.01895 and b=1.94411 so $p=\dfrac{11503}{1+e^{-.01895x+1.94411}}$. The graph of $p=\dfrac{11503}{1+e^{-.01895x+1.94411}}$ with the original four data points is displayed in Figure 2.

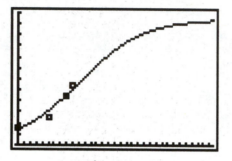

Figure 2

Some population projections based on this model are given in Table 3.

Year	x	World Population (millions)
1995	95	5338.5
2005	105	5882.7
2100	200	9934.5

Table 3

Procedures:
I.   The U.S. population in millions from 1910 to 1990 is given in Table 4 below (taken from <u>The World Almanac</u> 1993 by Funk and Wagnalls Corp.).

Year	U.S. Population (millions)	Year	U.S. Population
1910	92.20	1960	179.32
1920	106.02	1970	203.30
1930	123.20	1980	226.50
1940	132.16	1990	248.71
1950	151.33		

Table 4

1.   Graph a scatter plot of the data where x=0 corresponds to 1910. (See Appendix 1.5). Pick a time from this data where you think the growth rate is changing from increasing to decreasing. The time that you pick may not be obvious, so you may need to change your time estimate later to improve your model. Use this point to estimate L and then use Example 1 as a guide to fit a logistic model to the data. Record the equation of this model and your estimate of L. (Please note: this is different from example 1 since you can pick the inflection point directly from the given data points.)

2.   **G1** Print out the graph (see Appendix 4.1) of the logistic model overlaid on the scatter plot of the data in Table 4 (see Appendix 1.6).

II.   The national debt in billions of dollars is given in Table 5 below (taken from <u>The World Almanac</u> 1993 by Funk and Wagnalls Corp.).

Year	National Debt	Year	National Debt
1940	43.0	1970	370.1
1950	256.1	1980	907.7
1960	284.1	1990	3233.3

Table 5

1. Fit an exponential model of the form $y=ab^x$ to this data (see Appendix 7.7). Let x=0 correspond to the year 1940. Record your equation.

2.   Suppose Congress decides that in the year 1998, the growth rate of the national debt will be reversed to decreasing. Use the above exponential model to estimate the debt in 1998. Use this to estimate and record L for a logistic model.

3.   Use Example 1 as a guide to fit a logistic model to the data and include your estimate for 1998 with the original data. Record your logistic model.

4.   Predict the national debt in 2010 by the exponential and also the logistic model. Compute the difference between the two.

5.   Assume that the national debt follows this logistic model until 1998 and that the U.S. government does not borrow any more

money after 1998, but that each year the debt grows because of the accumulated interest. Suppose that the government is paying interest compounded annually at a 3% rate. To answer the following questions, find the amount of the national debt as a function of time where x=0 corresponds to 1940. This is to be a piecewise function where the first piece is the logistic growth function in problem 3 for the time period from 1940 to 1998. The second piece is to begin in 1998 with its value in 1998 the same as the value of the logistic growth model in 1998. For the period of time past 1998 assume that the growth of the debt is due only to the 3% annual interest rate. **G2** Print out the graph of this model and the graph of the logistic growth model (see Appendix 1.2 for graphing piecewise functions) overlaid on the scatter plot of the data, all on the same screen. In what year does the national debt, calculated by the "interest only" model, surpass that of the logistic model?

6.   In problem 3 you included the estimate for 1998 as part of the data to derive a logistic model. Derive a second logistic model without using the estimate for 1998 but using the same L. Record the equation corresponding to this model.

G3 Print the graph of this equation overlaid on a scatter plot of the data.

7.   Suppose you are to give a report about these two logistic models (problems 3 and 6) to a congressional budget subcommittee. Refer to the model in problem 3 as logistic model 1 and the model in problem 6 as logistic model 2. Respond to the anticipated questions and concerns of the committee members.
        a.   Which logistic model will have the smallest debt in the year 2000?

        b.   They want to know how the two logistic models are different. Look at the graphs of both logistic models and prepare a response.

        c.   They want you to explain the difference between the exponential model and the logistic model. Prepare an appropriate response.

        d.   Explain to the members of Congress the inaccuracy of using the exponential model to determine the limit, L, in problem 2, and what effect this will have on the resulting logistic model.

Checklist of calculator graph printouts and to be handed in:
☐   **G1**   Print out the graph of this model on the scatter plot of the data in Table 4.
☐   **G2**   Print out the graph of this model and the graph of the logistic growth model overlaid on the scatter plot of the data, all on the same screen.
☐   **G3**   Print out the graph of this equation overlaid on a scatter plot of the data.

Objectives:
1.   Given a set of data, choose an appropriate type of model from a linear, exponential or logistic model.
2.   Use curve fitting techniques to fit a set of data with an appropriate model.

Technology:
     TI-82, TI-85, or TI-92 graphing calculator.

Prerequisites:
     An introduction to curve fitting for linear, exponential and logistic models.

Overview: Table 1 contains the number of AIDS cases reported for each year from 1981 to 1992.[1] The actual number of new AIDS cases will be higher than the numbers in Table 1 since not all new AIDS cases are reported.

Year	Reported AIDS cases	Year	Reported AIDS cases
1981	199	1987	21,088
1982	744	1988	30,719
1983	2117	1989	33,595
1984	4445	1990	41,653
1985	8248	1991	43,701
1986	13,147	1992	45,472

Table 1

Procedures:
1.   Graph a scatter plot of the data in Table 1 (see Appendix 1.5). From this scatter plot, determine what you consider to be an appropriate type of model for the number of AIDS cases reported each year. Explain in some detail why you chose this model.

2.   Use the appropriate curve fitting technique to find and record the predicting equation for your model.

3.   **G1** Print out the graph of your model overlaid on a scatter plot of the data in Table 1.

4.   How many new AIDS cases does your model predict for the year 2010?

5.   Do you think that this model will still be valid in the year 2010? Explain your answer.

Checklist of calculator graphs to be handed in:
☐   **G1**   Print out the graph of your model overlaid on a scatter plot of the data in Table 1.

---
[1]From 1988 and 1994 Statistical Abstract

Objectives:
1. Develop a model to describe a logistic growth application for which we do not have symmetry about the point of inflection.
2. Linearize a data set to determine the parameters.
3. Use a piecewise function to describe a data set.

Technology:
TI-82, TI-85, or TI-92 graphing calculator

Prerequisites:
Introduction to logistic curves and to piecewise functions.

Overview: Although the logistic curve is a good model in many situations, the symmetry about the inflection point is a serious limitation. This symmetry implies that the forces producing the increase in growth rate are equal and opposite to the subsequent forces causing the reduction in growth rate. A more realistic asymmetrical model was proposed by Gilpin and Ayala. However, the formula they designed,

(1)
$$\frac{dN}{dt} = rN\left[1 - \left(\frac{N}{k}\right)^{\theta}\right]$$

contains a parameter, θ, which cannot be easily described by the physical problem. Because of the difficulties in using this model, it may be easier to create an asymmetrical logistic model by using two logistic curves of the form

(2)
$$N = \frac{K}{1 + e^{at+b}} + C$$

to form a piecewise function of the form

(3)
$$N = \begin{cases} \dfrac{K_1}{1 + e^{a_1 t + b_1}} + C_1 & if \quad t < t^* \\[4mm] \dfrac{K_2}{1 + e^{a_2 t + b_2}} + C_2 & if \quad t \geq t^* \end{cases}$$

.   A graphical interpretation of C_1, C_2, K_1, K_2, and t* is shown in Figure 1.

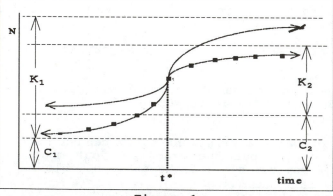

Figure 1

Notice that c_1 and c_2 are used to translate the functions vertically, while K_1 and K_2 correspond to the "height" of each function.

Procedures: The data in Table 1 represents the number of bacteria present in a culture at various times.

time (hours) t	Number of bacteria N
0	9
0.25	11
0.50	13
0.75	15
1.00	18
1.25	21
1.50	24
1.75	28
2.00	32
2.25	36
2.50	41
2.75	45
3.00	49
3.25	52
3.50	55
3.75	58
4.00	60
4.25	61
4.50	62
4.75	63
5.00	64
5.25	64
5.50	65

Table 1

A. Plot the data on your calculator with time as the independent variable.

B. Determine an appropriate value of time, t^*, which corresponds to the inflection point, and use t^* to break the data into two groups, one with $t \leq t^*$, the other with $t \geq t^*$ (the inflection point belongs to both portions.) Record your value for t^*.

C. Fit a logistic curve of the form (2) to the set of data where $t \leq t^*$ by linearizing the data and using linear regression. Refer to Figure 1 for estimating K_1 and c_1. Since c_1 is the height of a horizontal asymptote, it must be less than the smallest data value for N. (Hint: We can use the point of inflection and all points to the left of it to obtain an estimate for K_1.)

D.   Fit a logistic curve of the form (2) to the set of data where $t \geq t^*$. Refer to Figure 1 for estimating K_2 and c_2.

E.   Combine the two functions found in part C into a piecewise function of the form (3). Graph this function (see Appendix 1.2) on the same viewing screen as the original data (see Appendix 1.6). Record your function. Is the function a good model?

F.   Is your piecewise function continuous at the point of inflection, $t=t^*$? If you have not studied continuity yet, for now take continuity at $t=t^*$ to mean that the two pieces meet at $t=t^*$.

If not, translate one or both pieces of the function to make it continuous at $t=t^*$, while maintaining a reasonable model for the data. Record your function.

G.   **G1** Print a graph of this function (see Appendix 4.1) overlaid on a scatter plot of the data in Table 1.

Checklist of calculator graph printouts to be handed in:
☐   **G1**   Print a graph of this function overlaid on a scatter plot of the data in Table 1.

Objectives:
1.   Given the time intervals for an object to travel a fixed distance, perform the necessary calculations and graph a scatter plot of distance vs. time.
2.   Calculate average velocities and graph a scatter plot of velocity vs. time.
3.   Find a function which best fits the graph of distance vs. time.
4.   Write a program to calculate cumulative times and store the results in a list.

Technology:
     TI-82, TI-85, or TI-92 graphing calculator

Prerequisites:
1.   Find the derivative of a function or estimate the derivative from the graph of a numerical derivative.
2.   A brief introduction to programming.

Overview: A thin string with a weight attached is connected to a battery powered vehicle and placed over a pulley, as shown in Figure 1.

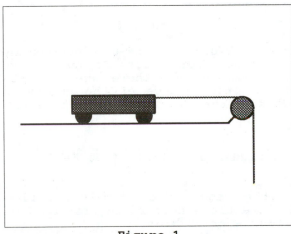

Figure 1

As the vehicle moves to the left, the pulley turns. The pulley has 10 spokes and as the pulley turns, each spoke interrupts a light beam and the time between interrupts is recorded. A computer or a CBL, a smart pulley and a propeller-driven "car" were used to collect such data. A sample run of data is given in Table 1.

Interrupts	1	2	3	4	5	6
Time	.1739	.1289	.1075	.0929	.0842	.0773

7	8	9	10	11	12	13	14
.0718	.0683	.0645	.0606	.0578	.0560	.0542	.0521

15	16	17	18	19	20	21	22
.0508	.0492	.0480	.0463	.0452	.0444	.0435	.0425

23	24	25	26	27	28	29	30
.0412	.0402	.0395	.0389	.0384	.0379	.0378	.0373

31	32	33	34	35	36
.0365	.0360	.0349	.0340	.0339	.0335

Table 1

The data recorded in Table 1 is the time between interruptions of the light beam. These interruptions are due to the spokes of the pulley as it turns. Since there are 10 interruptions per rotation and the circumference of the pulley is 15 cm., the distance traveled by the car per interruption is 1.5 cm.

Procedures:
I.
1. Store the sequence of times from Table 1 in list L_2 (see Appendix 7.1).

2. Generate the sequence of cumulative times up to each interrupt and store these numbers in list L_3. The calculations for the first four entries are:

$$L_3(1) = .1739$$
$$L_3(2) = .1739 + .1289$$
$$= L_3(1) + L_2(2)$$
$$L_3(3) = L_3(2) + L_2(3)$$
$$L_3(4) = L_3(3) + L_2(4)$$

As you can see, this is a laborious process. To simplify this task, write and execute a short program. You will probably use a "for" loop (see your calculator guidebook for programming guidelines) and the sum sequence command (see Appendix 9.2). Hand in a listing of your program.

3. Generate the sequence whose terms are the total distance traveled up to each time in list L_3 and store this sequence in list L_4. This task can be done with one sequence command. (see Appendices 9.1 and 6.2).

4. Generate a sequence of numbers whose terms are the average velocity (in cm/sec) for each of the time intervals in L_2. Store the terms of this sequence in L_5.

II.   The following problems involve calculator techniques of curve fitting (see Appendix 7.2).

1.   Fit a linear model to the time (L_3) and distance (L_4) data and record the resulting equation. **G1** Overlay the graph of the linear model on a scatter plot of time and distance data. Use the "eyeball" method to describe the fit. Store this graph as picture 1. (See Appendices 10.1 and 10.2) for details on storing and recalling pictures.)

2.   Fit an exponential model ($y=ab^x$) to the time and distance data and record the resulting equation. **G2** Overlay the graph of the exponential model on a scatter plot of time and distance data. Compare this fit with the one in part 1 (use the "eyeball" method.) Store this graph as picture 2.

3.   Fit a power model ($y=ax^b$) to the time and distance data and record the resulting equation. **G3** Overlay the graph of this model on a scatter plot of time and distance. Compare this fit with the best fit from 1 or 2. Store this graph as picture 3.

4.   Fit a polynomial model of degree three to the time and distance data and record the resulting equation. **G4** Overlay the graph of this model on a scatter plot of time and distance. Compare this fit with the fit of the models in parts 1-3. Store this graph as picture 4.

5.   Choose the two best fits in parts 1-4. **G5,G6** In each case overlay the graph of the derivative of the function on a scatter plot of time (L_3) and average velocity (L_5). Store these graphs as pictures 5 and 6.

6.   Choose and record the function which best fits the time and distance data and whose derivative best fits the time and average velocity data.

7.   Use a link to print out the graphs in parts 1-5 (see Appendix 4.1).

Checklist of calculator graph printouts to be handed in:
- ☐   **G1**   Overlay the graph of the linear model on a scatter plot of the time and distance data.
- ☐   **G2**   Overlay the graph of the exponential model on a scatter plot of the time and distance data.
- ☐   **G3**   Overlay the graph of this model on a scatter plot of the time and distance data.
- ☐   **G4**   Overlay the graph of this model on a scatter plot of the time and distance data.
- ☐   **G5,G6** In each case, overlay the graph of the derivative of the function on a scatter plot of time and average velocity.

APPENDICES

APPENDIX A Using the TI-82

1.1 FUNCTION GRAPHING

Walk-through example: Graph the function defined by $f(x)=2x^3-8x$

1. Display the MODE screen by pressing **[MODE]**.
2. Move the cursor down to Func and press **[ENTER]** to set the graphing mode.
3. Press **[Y=]** to display the Y= edit screen.
4. Use the arrow keys to move the cursor to one of the ten function positions such as Y2. Enter the following keystrokes to display $Y2=2x^3-8x$:
    [2] [X,T,θ] [∧] [3] [-] [8] [X,T,θ]
5. The "=" sign should be highlighted which indicates that the function at Y2 is selected for graphing. A function can be selected or unselected by placing the cursor on "=" and then pressing **[ENTER]**.
6. Define the viewing window by pressing **[ZOOM]** and then **[6]** (See Window Settings, Appendix 2.5).
7. You should see the graph of $y=2x^3-8x$ for the portion of the coordinate plane defined by the window settings. Your graph should look like Figure 1.
8. You can graph more than one function on the same screen by selecting several functions on the Y= edit screen.

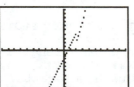

Figure 1

1.2 PIECEWISE FUNCTIONS

Walk-through example:
Graph $y=\begin{cases} 3x-2, & x<2 \\ x^2-2, & x\geq2 \end{cases}$

1. Press **[Y=]** to display the Y= edit screen. Place the cursor at Y1= and enter the following key strokes:
    [(] [3] [X,T,θ] [-] [2] [)] [(] [X,T,θ] [2ND] [MATH] [5] [2] [)] [+] [(] [X,T,θ] [x²] [-] [2] [)] [(] [X,T,θ] [2ND] [MATH] [4] [2] [)]
2. Define the viewing window by pressing **[ZOOM]** and selecting option 6 for the standard window. In some cases you may need to adjust your window variables to produce a better graph (See Appendix 2.5).
3. A piecewise function graph sometimes looks more realistic if the points are not connected. To set the dot mode rather than the connected mode, press **[MODE]**, move the cursor down to Connected, over to Dot and press **[ENTER]**. Press **[GRAPH]** and compare your graph with the one in Figure 2.

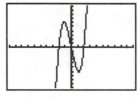

Figure 2

1.3 PARAMETRIC GRAPHING

Walk-through example:
Graph the function $y=2x^3-8x$ parametrically.

1. Change the form of this equation by letting x=T and $y=2T^3-8T$.
2. Change your calculator to the parametric mode by pressing **[MODE]**, move the cursor down to Func, move the cursor right to Par and press **[ENTER]**.
3. Press **[Y=]** to display the edit screen for parametric graphing. With the cursor on X_{1T}: Press **[X,T,θ]** and **[ENTER]**.
4. Your cursor should be at $Y_{1T}=$. Enter the following

keystrokes: [2] [X,T,θ] [∧] [3] [-] [8]
[X,T,θ].

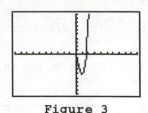

Figure 3

5. Press **[WINDOW]** to display the window
   edit screen for parametric graphing.
   Move the cursor down, type in the
   value, and press **[ENTER]** to complete
   each entry with the values below:

   Tmin=0   Xmin=-10   Ymin=-10
   Tmax=6   Xmax=10   Ymax=10
   Tstep=.1   Xscl=1   Yscl=1

6. Press **[GRAPH]** and compare your graph with the one in Figure
   3. In the parametric mode you can control the number of
   points that are plotted. For these settings sixty one
   points are plotted, the first one for T=0 then the values
   of T are incremented by .1 until T=6. For different graphs
   you will need to adjust the window values to get the
   desired graph.

1.4 POLAR GRAPHING

Walk-through example:
Graph r=4sin(2θ)

1. To set your calculator to the polar graphing and radian
   modes, press **[MODE]** to display the MODE menu. If the word
   radian is not highlighted, use the cursor keys to move the
   cursor to radian and press **[ENTER]**.

2. Move the cursor down to Func and right to Pol and press
   [ENTER].

3. Press **[Y=]** to display the edit screen for polar graphing.
   With the cursor at r1=, enter the following keystrokes:
   [4] [SIN] [(] [2] [X,T,θ] [)].

4. Press **[ZOOM]** and select Zstandard (#6)
   to set the window variables for polar
   graphing. This graph can be improved by
   changing the window settings.

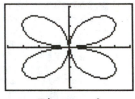

Figure 4

5. Press **[WINDOW]** and move the cursor to
   Xmin and make the following changes:
   Xmin=-4   Xmax=4
   Ymin=-4   Ymax=4

6. To display polar graphing coordinates of
   (r,θ) instead of rectangular coordinates of (x,y) during
   tracing, press **[WINDOW]**, move the cursor right to FORMAT,
   down to RectGC, right to PolarGC, and press **[ENTER]**.

7. Press **[GRAPH]** and compare your graph with the one in Figure
   4.

1.5 SCATTER PLOTS

The scatter plot feature is used to plot points from two lists of
the same length.
Walk-through example:
Graph a scatter plot of lists L1={.8, 1, 1.2, 1.5, 1.9} and
L2={.4, 1.1, 1.4, 2.3, 4} where the numbers in list L1 represent
the independent variable and those in L2 the dependent variable.

1. Press **[STAT]** to display the selection menu. Press **[1]** to
   select Edit.

2. You should see the columns headed by L1, L2 and L3. If
   there is data in L1 and L2, clear these two lists (See
   Appendix 6.1.3).

3. Enter the first set of data by placing the cursor in the L1
   column. Enter the first number of the set and press
   [ENTER]. Continue until all the data of the first set is
   entered.

4. Move the cursor to the column headed by L2. Enter the second set of data into list.
5. Press **[2nd] [Y=]** to display the STAT PLOTS screen. Move the cursor to Plot 1, Plot 2 or Plot 3 and press **[ENTER]**.
6. Select a Stat Plot by moving the cursor to the "on" position and pressing **[ENTER]**.
7. Move the cursor down to Type and with the cursor on the first type (points not connected), press **[ENTER]**.
8. Move the cursor down to Xlist and then to the list which corresponds to the independent variable (L1) and press **[ENTER]**.
9. Move the cursor down to Ylist and then right to the list which corresponds to the dependent variable (L2) and press **[ENTER]**. (If you had two variables p and t, where p=f(t), where the t values are stored in L5 and the p vales in L3, you would choose L5 for the Xlist and L3 for the Ylist.)
10. Move the cursor down to Mark, move it to □ and press **[ENTER]**.
11. To graph the scatter plot press **[ZOOM]** to display the zoom menu and select ZoomStat (option 9) and press **[ENTER]** if necessary. The scatter plot should appear on your graph screen. ZoomStat selects values for Xmin, Xmax, Ymin and

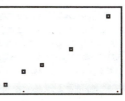

Figure 5

Ymax so that all data points appear on the graph screen. Compare your graph with the one in Figure 5. If other graphs appear, you should deselect functions on the y= menu or turn off other scatter plots.

1.6 GRAPHS OF FUNCTIONS OVERLAID ON A SCATTER PLOT

Sometimes you may want to compare the graph of a function stored on the **[Y=]** edit menu and a scatter plot.

1. Press **[Y=]**, enter Y1=3x-2 and turn off all other functions. A highlighted "=" sign indicates that a function is on. To turn it off, place the cursor on top of the "=" sign and press **[ENTER]**. Pressing **[ENTER]** again would turn it back on.

Figure 6

2. Follow the steps in Appendix 1.5 to graph the scatter plot and compare your graph with the one in Figure 6.
3. To turn off a scatter plot press **[2ND]** and **[Y=]** to display the STAT PLOTS screen. Move the cursor to the desired plot and press **[ENTER]**. Move the cursor to Off and press **[ENTER]**.

1.7 TRACE

The trace feature can be used in the function, parametric, polar or sequence mode.

Walk-through example: Graph and trace y=2x³-8x.

1. Follow the instructions in 1.1 to graph this function with a standard window.
2. To set the type of coordinates to be displayed when tracing, press **[WINDOW]**, and move the cursor to FORMAT. RectGC should be highlighted if you want to display rectangular coordinates. (To display polar coordinates when tracing a polar curve, move the cursor to PolarGC and press **[ENTER]**.)

3. Press **[TRACE]** to display the graphs and activate the tracer.
4. Notice the number in the upper right of the graph screen. A "1" indicates that the graph corresponding function y1 is displayed. A "2" indicates that the graph stored at y2= is displayed.
5. If more than one graph is present you can use the up or down arrow keys to select the graph that is to be traced.
6. The cursor is moved to points on the graph by using the right or left arrow key.
7. With the graph of $y=2x^3-8x$ displayed, move the cursor to a point on the graph whose displayed x-coordinate is .63829787. The y-coordinate should be -4.586267.
8. In this mode the calculator displays the coordinates of the point on the graph which is marked by the cursor.

2 Window Settings By changing the values of the window variables you can display the desired portion of the coordinate plane on the graph screen. You can also determine the values of the independent variable x which are chosen to plot the points of a graph. When you set values of Xmin and Xmax, the change in x, called Δx, is determined by the formula $\Delta x = \dfrac{Xmax - Xmin}{94}$. For a function $y=f(x)$ the first point on the graph is (Xmin, f(Xmin)), the second (Xmin+Δx, f(Xmin+Δx)), the third (Xmin+2Δx, f(Xmin+2Δx)) and the trend continues to the last point (Xmin+94Δx, f(Xmin+94Δx)), which is the same as (Xmax, f(Xmax)).

2.1 STANDARD WINDOW
The standard window is:

Xmin=-10	Xmax=10	Xscl=1
Ymin=-10	Ymax=10	Yscl=1

To obtain this window press **[ZOOM]** and then press **[6]** to select Zstandard. The graph window with the above window setting will be displayed.

2.2 TRIGONOMETRIC WINDOW (Radian Mode)
The Trig window has the following settings:

$$Xmin = -\left(\frac{47}{24}\right)\pi \qquad Ymin=-4$$

$$Xmax = \left(\frac{47}{24}\right)\pi \qquad Ymax=4$$

$$Xscl = \frac{\pi}{2} \qquad Yscl=1$$

These settings are chosen so that

$$\Delta x = \frac{\frac{47}{24}\pi - -\left(\frac{47}{24}\right)\pi}{94} = \frac{\frac{94}{24}\pi}{94} = \frac{\pi}{24}$$

This choice of Δx is "nice" for the special angles.
To obtain a Trig Window press **[ZOOM]** and then select ZTrig (option 7) and press **[ENTER]** (if necessary). The graph window with the above window settings will be displayed.

2.3 SQUARE WINDOW
Zsquare defines a window for which the ratio of (Xmax-Xmin):(Ymax-Ymin) is about 3:2. This window takes into consideration the size of the screen so that graphs will not look distorted. To change to a square window press **[ZOOM]** and then press **[5]**. The window variables will be changed and the current

graph will be displayed.

2.4 USER-DEFINED WINDOW
1.   Press **[WINDOW]** to display the window variables screen. Move the cursor down to the Xmin line, enter your choice of Xmin, and press **[ENTER]**.
2.   Enter your choice for Xmax and press **[ENTER]**.
3.   Continue until all values of the window variables are entered. You may return to the home screen by press **[2ND] [MODE]**.

2.5 FRIENDLY WINDOW (centered about 0)
A friendly window centered about 0 has

$$Xmax = -Xmin$$

$$\frac{2\,Xmax}{94} = \Delta x$$

Or Xmax=47Δx

For example if $\Delta x=.1$ then Xmax=4.7 and Xmin=-4.7. This window produces a graph whose chosen x coordinates are all numbers between -4.7 and 4.7 which differ by .1. For this setting the trace mode produces the Y values of a function for X values between -4.7 and 4.7 which differ by .1. Note: you don't input Δx directly, but when Xmin and Xmax are set, Δx is automatically set.

2.6 STAT WINDOW
To obtain a STAT window, press **[ZOOM]** and then select ZoomStat (option 9). ZoomStat defines a window so that all points of a scatter plot are displayed.

2.7 TURNING AXES OFF
To turn the coordinate axes off, press **[WINDOW]**, move the cursor right to FORMAT, down to AxesOn, right to AxesOff, and press **[ENTER]**. Press **[GRAPH]** to display the graph.

3.1 SET ZOOM FACTORS
The zoom factors determine the magnification when zooming in. They are initially set at 4 for XFact and Yfact. Both zoom factors are to be set at 10 for projects in this book.
1.   Press **[ZOOM]**, move the cursor to MEMORY and press **[4]** to select SetFactors.
2.   The cursor should be on the value of XFact. Type in 10 and press **[ENTER]**. With the cursor now on the value of YFact, type in 10 and press **[2ND] [MODE]** to quit.

3.2 ZOOM IN ON A POINT
1.   Select a function at the Y= menu and display its graph.
2.   Press **[ZOOM] [2]** to select Zoom In.
3.   Use the cursor keys (arrows) to move the cursor to a point which will be the center of the next viewing rectangle. This point is usually a point on the graph or close to a point on the graph.
4.   Press **[ENTER]** to display a "magnified" portion of the graph in a new viewing rectangle.
5.   You are still in the Zoom In mode. To Zoom In again move the cursor to the point which will be the center of the next viewing rectangle (if necessary) and press **[ENTER]**.
6.   You may continue the Zoom In process until you have a desired graph or you are beyond the range of your calculator.
7.   If you get a WINDOW RANGE ERROR press **[ENTER]** to quit. Now

press [ZOOM], move the cursor to memory, and select 1:ZPrevious. Your previous screen should be restored.

4 TI-Graph Link for the PC (Windows) If you have a Dos version of the Graph Link program, you will find instructions in your link manual.

4.1 PRINT A SCREEN IMAGE
1.   Boot up the Link-82 program on your computer.
2.   Display the calculator screen that you want to print.
3.   Firmly attach the PC link cable to your calculator.
4.   Select **Link** from your computer screen menu to display the LINK menu and choose "Get LCD from TI-82."
5.   When a submenu is displayed on the computer screen, choose Printer (small) or Printer (large).
6.   When the Receive box is displayed, select **Receive**.
7.   When the LCD save box is displayed you should see the calculator screen display:
          Save the LCD image to:
          PRINTER
      At this time select ok.
8.   When the Print box is displayed, you may select the number of copies if you want more than one. Select **OK** to print out your screen display.
(Use the HELP option for more information).

4.2 LOAD A TI-82 PROGRAM FROM THE COMPUTER
1.   Boot up the Link-82 program on your computer.
2.   Attach the PC link cable to your calculator.
3.   Press [2nd] [X,T,θ] on your calculator.
4.   Move the cursor to RECEIVE and press [ENTER]. You should see "Waiting".
5.   Select **LINK** and then **SEND** from your computer screen menu.
6.   When the send box is displayed, select the appropriate drive and directory.
7.   Select **OK** to display a list of TI-82 file names on disk. Select a file name (press **CTRL** and select another file name if more than one file is to be sent).
8.   At the time select **OK** to transmit these files.
9.   If a file is already stored in your calculator, you will have the option to:
          1. Rename
          2. Overwrite
          3. Omit
          4. Quit
10.   Press the number of your choice. If you choose "1. Rename", you will be prompted for a new name.

4.3 RUN A PROGRAM
1.   The program must be stored in your calculator.   (See Appendix 4.2).
2.   Press [PRGM] to display the names of the programs stored in your calculator.
3.   Move the cursor to the name of the program that you want to run and press [ENTER] to mark this program.
4.   If you do not see the name, move the cursor down to see names not on the first screen. If the name does not appear, you should store the program in your calculator. (See Appendix 4.2).
5.   Press [ENTER] to display the program name on the HOME screen.

6.   Press **[ENTER]** to run the program.
7.   If you get an error message while running a program you should press **[2]** to quit.
8.   You can stop the execution of a program by pressing **[ON]**. If you do, press **[2]** to quit when you see ERR:BREAK.

5.1 TABLES   The TABLE option is usually used to display the values of functions which are selected for graphing. It may also be used to compare function values of two or more functions. Walk-through example: Display the coordinates of $y=2x^3-8x$ by using the table option.
1.   Go through the instructions in Appendix 1.1 to graph this function with a standard window.
2.   Press **[2ND] [WINDOW]** to display the TABLE SETUP screen. The value of TblMin is the initial choice of the independent variable for evaluating one or more functions. The value of ΔTbl determines the increment for the independent variable. If ΔTbl=1 then the second value of the independent variable is TblMin +1, the third value is TblMin +2 etc.
3.   If you choose Auto for the independent and the dependent variables, a table of function values will be generated for the values of **x** defined by TblMin and ΔTbl.
4.   Enter the following values for TABLE SETUP:
          TblMin=0
          ΔTbl=1
          Indpnt: Auto
          Depend: Auto
5.   Press **[2nd] [GRAPH]** to display the table and compare your results with the partial listing below:

X	Y1
0	0
1	-6
2	0
3	30

6.   Many situations call for the user to choose the value of the independent variable. Press **[2nd] [WINDOW]**. Move the cursor down to Indpnt:, move the cursor to Ask and press **[ENTER]**.
7.   Press **[2nd] [GRAPH]** to display the Table. With the cursor under X type 2.1 and press **[ENTER]**. You should see 1.722 in the Y column.
8.   If you have selected several functions for graphing, their function values will appear in the table. You only see one column of x values and two columns of function values on a table screen. If more than two functions are selected, press the right arrow key to view values of other functions.

6.1.1 STORE DATA IN TWO LISTS
Walk-through example: Store {.8, 1, 1.2, 1.5, 1.9} in list L1 and {.4, 1.1, 1.4, 2.3, 4} in list L2.
1.   Press **[STAT]** to display a selection menu.
2.   Select 1:Edit to display lists L1, L2 and L3. Use the right arrow key to display lists L4, L5 and L6.

TI-82

3.   If lists L1 and L2 contain data, use the instructions in
        Appendix 6.1.2 to clear these lists.
4.   Use the arrow keys to place the cursor on the first
        position of list L1.
5.   Enter data into the list by entering the first number of
        the list and moving down one position.
6.   Continue entering numbers until you have entered all of the
        data.
7.   Enter the above data in list L2 in a similar manner.
8.   Press **[2nd] [MODE]** to exit this mode.
9.   From the home screen press **[2nd] [1] [ENTER]** to display the
        contents of L1. Check the list with the one in the walk
        through example.
10.   From the home screen press **[2nd] [2]** to check the contents
        of L2. Notice that the displayed numbers are separated by
        spaces rather than commas.

6.1.2 CLEAR THE CONTENTS OF A LIST
1.   Press **[STAT]** and select Clrlist by typing 4 or placing the
        cursor on 4 and pressing **[ENTER]**.
2.   Choose the appropriate list name by pressing **[2nd]** and a
        numeric key from 1 to 6 to name L1 to L6. Press **[ENTER]**.
        You can clear more than one list by choosing list names
        separated by a comma.

6.2 THE ALGEBRA OF LISTS
Walk-through example:
1.   Store the numbers 1, 2, 3, 4, 5 in list L1 and return to
        the home screen by pressing **[2ND] [MODE]**.
2.   To double each number in list L1 and store the results in
        list L2, enter the following keystrokes from the home
        screen:
        [2] [2ND] [1] [STO▸] [2ND] [2] [ENTER]
        The set {2,4,6,8,10} should be displayed.
3.   The Algebra of lists can be extended to more complicated
        expressions such as $\ln(5x^2+4x+7)$. To apply this function to
        the numbers in the original list L1 and store the results
        in L3, first store the numbers 1, 2, 3, 4, 5 in L1. Enter
        the following keystrokes on the home screen.
        **[1n] [(] [5] [2ND] [1] [∧] [2] [+] [4] [2ND] [1] [+] [7]
        [)] [STO▸] [2ND] [2] [ENTER]**.
        You should see a list of 5 numbers. The first number in L3
        is 2.772588722. Move the cursor to the right to see the
        others.

7 Curve Fitting (Regression) A set of data must be stored in two
lists of equal length before you can use your calculator to fit
a curve to the data.

7.1 STORE DATA PAIRS IN TWO LISTS
Walk-through example:
Store data points in two lists.   L1={.8, 1, 1.2, 1.5, 1.9} and
L2={.4, 1.1, 1.4, 2.3, 4} where the numbers in list L1 represent
the independent variable and those in L2 the dependent variable.
Follow the instructions in Appendix 6.1.1.

7.2 LINEAR REGRESSION
1.   First enter your data in two lists, say L1 and L2. Press
        [STAT], move the cursor right to CALC and press **[5]** to
        choose LinReg from the STAT CALC menu.   When LinReg(ax+b)
        appears on the screen, type **[2nd] [1] [,] [2nd] [2]** and

press **[ENTER]**. Calculator will perform linear regression with the first list, L1, as the independent variable, x.

2. Your display should read:

y=ax+b
a=3.173796791
b=-2.222459893
r=.9897152859

The r-value is called the correlation coefficient and is a measure of how well the line fits the data.
Note: Option 9 uses linear regression to fit a line to y=a+bx. This is similar to option 5 except that the parameters a and b are switched.

7.3 QUADRATIC REGRESSION
1. Store the data from Appendix 7.1 in L1 and L2.
2. Press **[STAT]**, move the cursor right to CALC and press **[6]** to select QuadReg from the STAT CALC menu. When QuadReg appears on the screen press **[2nd]** **[1]** **[,]** **[2nd]** **[2]** **[ENTER]**. The calculator will perform quadratic regression with the first list, L1, as the independent variable, x.
3. Your display should read

$y=ax^2+bx+c$
a=1.157726246
b=.03060099
c=-.2691837943

7.4 CUBIC REGRESSION
Proceed as in Appendix 7.3 but select option 7:CubicReg instead of 6:QuadReg. This algorithm uses regression to find a, b, c, d which defines the "best" cubic equation of the form $y=ax^3+bx^2+cx+d$ to fit the data in the two specified lists. CubicReg L1, L2, for example, will treat L1 as the independent variable, x.

7.5 QUARTIC REGRESSION
Proceed as in Appendix 7.3 but select option 8:QuartReg instead of 6:QuadReg. This algorithm uses regression to calculate values of a, b, c, d and e which define an equation of the form $y=ax^4+bx^3+cx^2+dx+e$ which best fits the data in the two specified lists. QuartReg L1, L2, for example, will treat L1 as the independent variable, x.

7.6 LOGARITHMIC REGRESSION
Proceed as in Appendix 7.3 but select option 0:LnReg instead of 6:QuadReg. This algorithm uses regression to fit a curve of the form y=a+blnx to data in two specified lists. LnReg L1, L2, for example, will treat L1 as the independent variable, x.

7.7 EXPONENTIAL REGRESSION
Proceed as in Appendix 7.3 but select option A:ExpReg instead of 6:QuadReg. This algorithm uses regression techniques to fit a curve of the form $y=ab^x$ to a set of data in two specified lists. ExpReg L1, L2, for example, will treat L1 as the independent variable, x.

7.8 POWER REGRESSION
Proceed as in Appendix 7.3 but select option B:PwrReg instead of 6:QuadReg. This algorithm uses regression techniques to fit a curve of the form $y=ax^b$ to data in two specified lists. PwrReg L1, L2, for example, will treat L1 as the independent variable, x. Notice that the curves in options A and B are different types.

TI-82

7.9.1 STORE A REGRESSION EQUATION ON THE Y= EDIT MENU

Walk-through example: Store the linear regression equation for the sets in Appendix 7.1.

1. Go through the curve fitting process outlined in Appendix 7.2.
2. Press [y=] to display the y= edit menu and move the cursor to the right of y1= or to the right of any y= position which has no function currently stored.
3. Press [VARS] to display the VARS menu and press [5] to choose Statistics. Move the cursor right to EQ and press [7] (RegEQ).
4. The y= edit menu should be displayed and the regression equation stored and selected for graphing. You should see y=3.1737967914439x+-2.2224598930482.

7.9.2 GRAPH THE REGRESSION EQUATION

Walk through example: Graph the regression equation which was stored in Appendix 7.9.1.

At this stage one normally overlays the graph of the regression equation on a scatter plot of the data.

1. Press [y=] to display the y= edit screen and make sure that the regression equation is the only one highlighted.
2. Press [2ND] [y=] to display the STAT PLOTS edit screen.
3. Press [1] to select Plot 1.
4. Position the cursor on "on" and press [ENTER].
5. Move the cursor down to Type and with the cursor on the first type press [ENTER].
6. Move the cursor to L1 on the xList and press [ENTER] to select this list to represent the independent variable.
7. Move the cursor to L2 on the ylist and press [ENTER] to select this list to represent the dependent variable.
8. Move the cursor to any of the three characters on Mark and press [ENTER] to select the symbol that will be used to represent a data point.
9. Press [ZOOM] and press [9] (ZoomStat) to set the window variables and display a graph of the regression equation overlaid on a scatter plot of the data in lists L1 and L2. Your graph should look like the one in Figure 1.

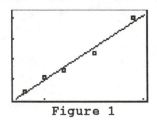

Figure 1

8.1 NDERIV (Numerical Derivative)

1. Press [MATH] to display the math operations menu and press 8 to choose nDeriv.
2. The syntax for the nDeriv command is: nDeriv (f(x),x,a). For example nDeriv ($3x^2+x$,x,2) returns an approximation to f'(x) at x=2 where f(x)=$3x^2+x$. With "nDeriv(" displayed, enter the following keystrokes [3] [X,T,θ] [x^2] [+] [X,T,θ] [,] [X,T,θ] [,] [2] [)] [ENTER]. After completing the nDeriv command you should see 13.
3. If you already have the function stored on the y= menu, say at y3, you may enter nDeriv(y3,x,2).

8.2 $\frac{dy}{dx}$ (Numerical Derivative) from the graph of y=f(x)

Walk-through example: Graph $y=3x^2+x$ with a standard window and follow the directions below to approximate the slope of the tangent line to this graph at x=1.0638298.

1. Press **[2ND] [TRACE]** to display the CALCULATE menu.
2. Choose dy/dx and the current graph will be displayed. (You need at least one graph activated on the Y= menu).
3. If you have more than one graph displayed choose the desired one by using the up or down arrow **[▲]** or **[▼]**.
4. Use the right or left arrow **[▶]** or **[◀]** to trace to the point where x=1.0638298 and press **[ENTER]**.
5. The numerical derivative of the function at this value of x is 7.3829787.

8.3 $\frac{dr}{d\theta}$

Walk through example: Approximate $\frac{dr}{d\theta}$ at $\frac{\pi}{6} \approx .523$ for r=4sin(2θ).

1. Follow the directions with the specified window settings in 1.4 to graph r=4sin(2θ).
2. Press **[WINDOW]**, move the cursor to FORMAT and press **[ENTER]**.
3. Position the cursor over PolarCG and press **[ENTER]**.
4. Press **[GRAPH]** to display the graph of r=4sin(2θ), and press **[2ND] [CALC]** to display the CALCULATE menu.
5. Press **[3]** to select $\frac{dr}{d\theta}$ and display the graph of r=4sin(2θ).
6. Move the cursor to the point whose θ-coordinate is about .523 and press **[ENTER]**.
7. You should see dr/dθ=3.9999973 which can probably be interpreted as 4.

8.4 FNINT (Numerical Integral)

1. Walk-through example: Approximate $\int_{1}^{3} x^3\, dx$.

2. Press **[MATH]** to display the math operations menu and press **[9]** to choose fnInt under the MATH option.
3. The syntax for fnInt is fnInt(f(x),x,a,b). For example, fnInt(x^3,x,1,3) returns an approximation of the integral $\int_{1}^{3} x^3\, dx$. After completing the fnInt command, press **[ENTER]** to display 20, which is the approximation of $\int_{1}^{3} x^3\, dx$.

8.5 $\int f(x)\, dx$ (Numerical Integral) from the graph of y=f(x)

Walk through example: Approximate $\int_{1}^{3} x^3\, dx$.

1. Store $y=x^3$ on the y= menu and highlight the function for graphing.
2. Set the window variables as:
   xmin=-4.7   ymin=0
   xmax=4.7   ymax=30
   xscl=1   yscl=2

3. Press, [2ND] [TRACE], to display the calculate menu.
4. Press [7] and the current graph is displayed. (You need at least one function activated on the y= menu)
5. If you have more than one graph displayed, choose the desired one by using the up or down arrow [▲] or [▼].
6. When asked for "lower limit", move the cursor to the point whose x-coordinate is 1 and press [ENTER].
7. When asked for "upper limit" move the cursor to the point whose x-coordinate is 3 and press [ENTER].
8. The desired region is shaded and an approximation of 20 for $\int_{1}^{3} x^3\, dx$ is displayed.

8.6 DRAW A TANGENT LINE TO A POLAR GRAPH

Walk-through example: Draw a line tangent to the graph of $r=4\sin(2\theta)$ at the point where $\theta=\frac{\pi}{6}\approx.523$.

1. Follow the instructions in Appendix 1.4 to graph $r=4\sin(2\theta)$.
2. Press [2ND] [PRGM] [5] to select Tangent(and display the graph of $r=4\sin(2\theta)$.
3. Use the right arrow key to move the cursor to the point whose θ coordinate is .5235987756 ($\frac{\pi}{6}\approx.5235987756$) and press [ENTER].
4. Your graph should look similar to the one in Figure 1.

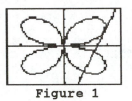

Figure 1

9.1 THE SEQUENCE COMMAND

The seq(command is used for evaluating selected terms of a sequence or storing selected terms of a sequence in a list.
Walk-through example: Display the 3rd, 5th, 7th, 9th and 11th term of the sequence $a_n=n^3-2$.

1. Press [2ND] [STAT] to display the list menu and press [5].
2. "Seq(" with a blinking cursor should be displayed on the home screen. This is a prompt to enter the parameters of this command. With the cursor at this position enter the following key strokes:
    [ALPHA] [LOG] [∧] [3] [−] [2] [,] [ALPHA] [LOG] [,] [3] [,] [1] [1] [,] [2] [)].
    You should see: seq(N∧3-2,N,3,11,2).
    $a_N=N\wedge3-2$ defines the sequence $\{a_N\}$, N represents the variable, 3 the first term to be evaluated, 11 the last term to be evaluated and 2 is the increment. The sequence command with these instructions evaluates a_3, a_5, a_7, a_9 and a_{11}. Press [ENTER] and you should see: {25 123 341 727...}. Press the right arrow key to see the value of a_{11} which is 1329.

9.2 THE SUM COMMAND

Walk-through example: Calculate $\sum_{N=1}^{5} \frac{1}{N}=1+\frac{1}{2}+\frac{1}{3}+\frac{1}{4}+\frac{1}{5}$.

1.   Press **[2ND] [STAT]** and move the cursor right to MATH to
       display the LIST/MATH menu.
2.   Press **[5] [2ND] [STAT] [5]** to display Sum Seq(.
3.   With the cursor just to the right of Sum Seq(enter the
       following keystrokes:
       **[1] [÷] [ALPHA] [LOG] [,] [ALPHA] [LOG] [,] [1] [,] [5]
       [,] [1] [)] [ENTER]**
4.   The sum of 2.283333333 should be displayed.

9.3 GRAPH SELECTED TERMS OF A SEQUENCE

Walk-through example: Graph the first 20 terms of the sequence
defined by $a_n = 2n - 5$.

1.   Press **[MODE]**, move the cursor down to Func. Use the
       right arrow key to move the cursor to Seq and press
       [ENTER] to activate the sequence graphing mode.
2.   Move the cursor to Dot and press **[ENTER]**.
3.   Press **[Y=]** to display the sequence edit screen.
4.   With the cursor at $u_n=$ enter the keystrokes: **[2] [2ND]
       [9] [-] [5] [ENTER]**.
5.   Press **[WINDOW]** to display the window variables for
       sequence graphing.
6.   Move the cursor down to UnStart, enter the value of the
       first term of the sequence which is -3, and press
       [ENTER]. At this time move the cursor to nStart since we
       are not graphing the sequence $\{v_n\}$.
7.   Enter the values of the other variables as follows:

nStart=1	Xmin=0	Ymin=-1
nMin=1	Xmax=20	Ymax=40
nMax=20	Xscl=1	Yscl=2

8.   Press **[GRAPH]** to display the graph of the first 20 terms
       of the sequence $a_n = 2n - 5$.
9.   Press **[TRACE]** and move the cursor to the point where
       n=16. The value of u_n should be 27. If this is not the
       case, repeat the above instructions.

9.4 GRAPH SELECTED TERMS OF A RECURSIVE SEQUENCE

Walk-through example: Graph the first 20 terms of the sequence
defined by $a_1 = 2$ and $a_n = a_{n-1} + \dfrac{1}{n}$

1.   Press **[MODE]** to display a selection menu. Use the arrow
       keys to move the cursor to Seq and press **[ENTER]** to
       activate the sequence mode for graphing.
2.   Use the arrow keys to move the cursor to Dot and press
       [ENTER].
3.   Press **[Y=]** to display the sequence edit screen.
4.   At $u_n=$ press **[2nd] [7] [+] [1] [÷] [2nd] [9]** to display
       $u_n = u_{n-1} + \dfrac{1}{n}$

5.   Press **[WINDOW]** and choose the following settings:

u_nStart=2	Xmin=0	Ymin=0
nStart=1	Xmax=20	Ymax=5
nMin=1	Xscl=2	Yscl=2
nMax=20		

       u_nStart is the value of the nth term when n= nStart.
       nStart is the value of n at which calculations begin and
       should be 1 when a sequence is defined recursively. nMin
       is the number of the first term to be graphed. nMax is
       the number of the last term to be graphed.
6.   Press **[GRAPH]** to display the desired graph.
7.   Trace to evaluate the terms.

Compare your results with the following: $u_1=2$, $u_4 = 3.45$, $u_{15}=4.318229$. If your results are different, check the formula at $u_n=$ and check your window settings.

10.1 STORE A PICTURE

Walk-through example: Store the graph of $y=x^2$ as a picture.

1. Store $y=x^2$ at y1=, set a standard window (Zoom 6), and display the graph of $y=x^2$. If other functions are graphed, go back to the y= Editor and deselect them. (See Appendix 1.1)
2. Press **[2ND] [PRGM]**, move the cursor to the right to STO, and press **[1]** (StorePic).
3. You should see StorePic on the home screen. To store the graph an appropriate Pic name must be entered. Press **[VARS] [4]** to display the six Pic names.
4. Press **[1]** and **[ENTER]** to store this graph as Pic1 and display the graph.
5. If you have a picture named Pic 1, it will be replaced by this one.

10.2 RECALL A PICTURE

Walk-through example: Recall the picture that was stored in Appendix 10.1.

1. Press **[Y=]** and deselect all functions for graphing. Turn the axes off by pressing **[WINDOW]** and moving the cursor fright to FORMAT, down to AxesOn, over to AxesOff, and pressing **[ENTER]**. Press **[GRAPH]**. Your graph screen should be clear.
2. Press **[2ND] [PGRM]** and move the cursor right to STO and press **[2]** (RecallPic). You should see "RecallPic" on the screen.
3. Press **[VARS] [4]** to select the list of Pic names. Press **[1] [ENTER]**. The graph of $y=x^2$ should appear on the graph screen.
4. If you get ERR:UNDEFINED, you did not have a picture stored with the Pic name that you chose in 3 above.
5. If you get this error, press **[1]** to go to the previous command which is RecallPic. At this time you can choose a new Pic name or press **[CLEAR]** to clear this command.
6. To check to see which names are used, press **[2ND] [+]** and press **[2]** (Delete). YOU MUST BE CAREFUL OR YOU WILL ACCIDENTLY DELETE INFORMATION FROM YOUR CALCULATOR!
7. A list of categories should be displayed. Press **[7]** (Pic) to see a list of all Pic names that are active. Press **[2ND] [MODE]** to QUIT and return to the home screen.
8. After you are finished with the picture, you can erase it by pressing **[2nd] [PRGM] [1] [ENTER]**. This will paste the command ClrDraw on the home screen. Press **[ENTER]** again to execute this command.

10.3 OVERLAY A PICTURE ON A GRAPH

Walk-through example: Overlay the picture which was stored in Appendix 10.1 on the graph of $y=(.3)^x$.

1. Store the graph of $y=x^2$ with a standard window as a picture with name Pic 1. (See Appendix 10.1 above).

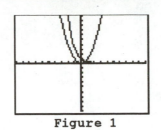

Figure 1

2. Enter y=(.3)^x on the Y= menu. Display its graph with a standard window.
3. Recall the picture named Pic 1. (See Appendix 10.2 above).
4. Your graph should look like the one in Figure 1.
5. In order to accurately compare the two functions, the windows for the current graph and the picture must be the same. If your graph appears to have two sets of axes, compare the window.

11.1 APPROXIMATE A SOLUTION TO AN EQUATION
Walk-through example: Solve cosx=x-1, -10≤x≤10.
1. Write the equation as cosx-x+1=0.
2. Press **[y=]** to display the y= edit screen.
3. Store y=cosx-x+1 at a y= position and deselect any other functions which may be stored.
4. Press **[ZOOM] [6]** to set up a standard graphing window and display the graph.
5. Press **[2ND] [TRACE]** to display the CALCULATE menu and press **[2]** to choose root.
6. You should see the prompt "Lower Bound?". Position the cursor to the left of the x-intercept and press **[ENTER]**.
7. You should now see the prompt "Upper Bound?". Move the cursor to the right of the x-intercept and press **[ENTER]**.
8. The prompt "Guess?" should be displayed. Position the cursor so that it is close to the x-intercept and then press **[ENTER]**. The solution, x=1.2834287, should be displayed on the screen. (You can select the number of digits displayed by selecting Float on the MODE menu.)

11.2 APPROXIMATE A SOLUTION TO A SYSTEM OF TWO EQUATIONS
Walk-through example: Solve y=cosx and y=x-1 for -10≤x≤10.
1. Write the equation cosx=x-1 and follow A11.1.

1.1 FUNCTION GRAPHING

Walk-through example: Graph $y = 2x^3 - 8x$.

1. Display the mode screen by pressing [2ND] [MORE].
2. Move the cursor down to Func and with the cursor on Func press [ENTER] to set the graphing mode. Press [EXIT] to return to the home screen.
3. Press [GRAPH] [F1] to display the y= edit screen.
4. Use the down arrow key to move to one of the function positions. If you are at a position where an unwanted function is stored, press [CLEAR] to clear the function storage position.
5. Enter the following keystrokes to store this function at y1: [2] [X-VAR] [^] [3] [-] [8] [X-VAR].
6. The "=" sign should be highlighted in order to graph this function. To select or unselect a function for graphing the cursor should be on any character in the function definition. If SELCT is above [F5] press [F5]. If GRAPH is directly above [F5] and SELCT is not visible, press [F1] and then [F5]. Pressing SELCT again would unselect the function.
7. Display the GRAPH menu (you may need to press [EXIT] if you are at a sub menu) and press [F3] to display the Zoom menu. Press [F4] to choose ZSTD on the Zoom menu. ZSTD sets up a standard viewing rectangle (See Appendix 2.1).
8. Your graph should look like the one in Figure 1.
9. You can graph more than one function on the same screen by selecting several functions on the y= edit screen, using [F5] (SELECT).

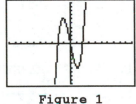

Figure 1

1.2 PIECEWISE FUNCTIONS

Walk-through example: Graph $y = \begin{cases} 3x-2, & x<2 \\ x^2-2, & x\geq2 \end{cases}$

1. Press [GRAPH] [F1] to display the y= editor. Place the cursor at y1= and enter the following keystrokes:
    [(] [3] [X-VAR] [-] [2] [)] [(] [X-VAR] [2ND] [2] [F2] [2] [)] [+] [(] [X-VAR] [x²] [-] [2] [)] [(] [X-VAR] [F5] [2] [)]
2. Turn off all other functions that may be selected for graphing.
3. Define the viewing rectangle by pressing [EXIT] [EXIT] to return to the GRAPH menu and then press [F3] [ZOOM].
4. Press [F4] to select the standard viewing rectangle. In some cases you may want to adjust the RANGE variables to produce a better graph (See Appendix 2.4).
5. A piecewise function graph is sometimes misleading in the connected graphing mode. To set the Drawdot Mode, the GRAPH menu is displayed then press [MORE] [F3] to display the FORMT menu. Move the cursor down to Drawline and over to Drawdot and press [ENTER] to highlight the Drawdot mode. Press [F5] to display a graph which should look like the one in Figure 2.

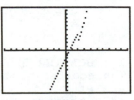

Figure 2

1.3 PARAMETRIC GRAPHING
Walk-through example.
Graph the function $y=2x^3-8x$ parametrically.
1.   Change the form of this equation by letting x=t and $y=2t^3-8t$.
2.   Change your calculator to the parametric mode by pressing [2ND] [MORE]. Move the cursor down to Func, over to Param and press [ENTER].
3.   Press [GRAPH] [F1] to select the E(t)= edit screen.
4.   With the cursor at xt1= press [F1] and [ENTER].
5.   With the cursor at yt1=, enter the following keystrokes: [2] [F1] [∧] [3] [-] [8] [F1] [ENTER].
6.   The "=" sign at xt1 and yt1 should be highlighted. If this is not the case, place the cursor on any character on the xt1 or the yt1 line and press [F5] to select the function for graphing.
7.   Press [EXIT] [F2] to display the RANGE editor screen for parametric graphing. Type in each value below, pressing [ENTER] after each entry.

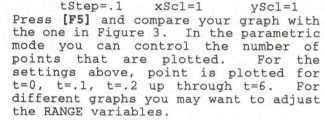

tMin=0	xMin=-10	yMin=-10
tMax=6	xMax=10	yMax=10
tStep=.1	xScl=1	yScl=1

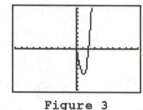

8.   Press [F5] and compare your graph with the one in Figure 3. In the parametric mode you can control the number of points that are plotted. For the settings above, point is plotted for t=0, t=.1, t=.2 up through t=6. For different graphs you may want to adjust the RANGE variables.

Figure 3

1.4 POLAR GRAPHING
Walk-through example: Graph $r=4\sin(2\theta)$
1.   To set your calculator to the polar graphing and radian modes: press [2ND] [MORE], move down to Radian and press [ENTER]. Move the cursor down to Func over to Pol and press [ENTER].
2.   Press [GRAPH] to display the polar graph menu and [F1] to display the r(θ)= editor screen.
3.   With the cursor at r1= enter the following key strokes: [4] [SIN] [(] [2] [F1] [)].
4.   Press [EXIT] and [F2] to display the RANGE editor screen for polar graphing. Type in each value below, pressing [ENTER] after each entry.

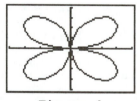

θMin=0	xMin=-4	yMin=-4
θMax=2π	xMax=4	yMax=4
θStep=π/24	xScl=1	yScl=1

5.   Press [F5] to display a graph. Your graph should look like the one in Figure 4.

Figure 4

1.5 SCATTER PLOTS
The scatter plot feature is used to plot points whose coordinates are stored in two lists of the same length.
Walk-through example:
Graph a scatter plot of the lists L1={.8,1,1.2,1.5,1.9} and L2={.4,1.1,1.4,2.3,4} where the numbers in list L1 represent the independent variable and those in L2 the dependent variable. We will first store the above data in lists L1 and L2. If lists L1

and L2 contain unwanted data, clear them (See Appendix 6.1.2).

1.   Press **[STAT]** to display the STAT menu. Press **[F2]** to display the stat editor.

2.   At the xlist line, press **[ALPHA] [7] [1]** to name this list as L1 and press **[ENTER]**.

3.   With the cursor on the ylist line, press **[ALPHA] [7] [2]** to name this list as L2 and press **[ENTER]**.

4.   You are now ready to enter the elements of lists L1 and L2. Type the first element of L1 which is .8 and press **[ENTER]**.

5.   With the cursor at y1, type over the "1" with .4 and press **[ENTER]**.

6.   Type the value of x2 press **[ENTER]**, the value of y2 and press **[ENTER]**. Continue until you have entered all values through x5 and y5.

7.   To check your results press **[EXIT]** twice to exit the stat mode and press **[2nd] [-]** to display the LIST menu. Press **[F3]** to display list names.

8.   You should see the names L1 and L2. If not, press **[MORE]** until you do. Press the function key directly below L1 and then **[ENTER]**. You should see {.8 1 1.2 1.5 1.9}.

9.   Press the function key directly below L2 and **[ENTER]** to check the numbers in list L2. Press **[EXIT]** twice.

10. Set the graphing mode to Func (See Appendix 1.1).

11. To set the range variables press **[GRAPH] [F2]** and type in the values below.

    xMin=0   yMin=0
    xMax=2   yMax=4.5
    xScl=.1   yScl=.1

12. Press **[STAT] [F2]** to display the current lists names. L1 and L2 should be the xlist and ylist names. If not select them at this time, by pressing the appropriate function key.

13. Press **[EXIT] [F3]** to display the STAT DRAW menu and press **[F2]** (SCAT) to display the scatter plot. Press **[CLEAR]** to remove the menu from the GRAPH screen. Your graph should look like the one in Figure 5.

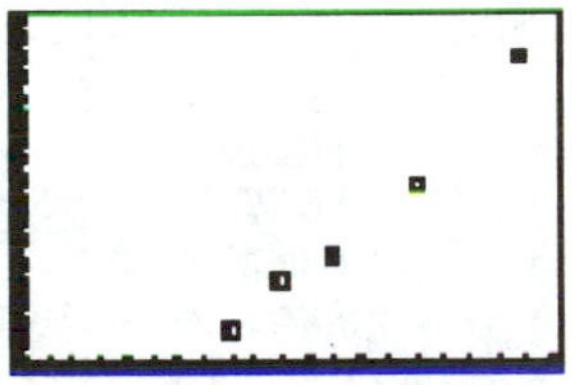

Figure 5

1.6 GRAPHS OF FUNCTIONS OVERLAID ON A SCATTER PLOT

You may want to compare the graph of a function stored on the y= edit menu and the scatter plot in Appendix 1.5.

1.   Press **[GRAPH] [F1]** to display the y(x)= edit menu. At y1= type the following keystrokes **[3] [X-VAR] [-] [2]**. Make sure that this function is the only function selected for graphing.

2.   Press **[STAT] [F2]** and select L1 for Xlist and L2 for the Ylist. Press **[EXIT] [F3] [F2]** and **[CLEAR]**. Compare your graph with the one in Figure 6.

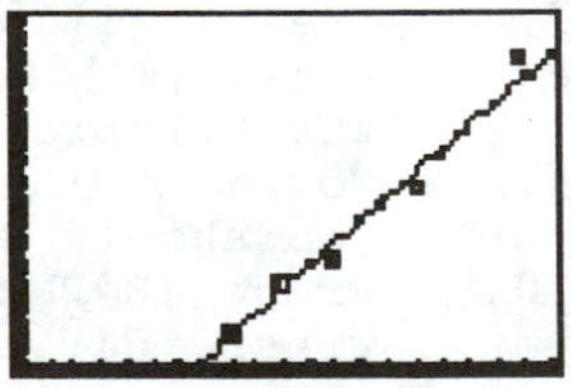

Figure 6

1.7.1 TRACE A FUNCTION GRAPH

The trace feature can be used in the function, polar, parametric or differential equation mode.

Walk-through example: Graph and trace $y=2x^3-8x$.

1.   Follow steps 1-9 in Appendix 1.1 to graph this function with a standard viewing rectangle.

2.   To set the type of coordinates to be displayed press **[GRAPH} [MORE] [F3]** to display the FORMAT screen. If you

want to display rectangular coordinates, RectGC should be highlighted. To display polar coordinates when tracing a polar graph, move the cursor to PolarGC and press [ENTER].

3. Press [F4] to activate the tracer.
4. Notice the number in the upper right corner of the screen. The number 1 indicates that the graph stored at y1= is displayed. The number 2 indicates that the graph of the function stored at y2= is displayed.
5. If more than one graph is displayed, you can use the up or down arrow keys to select the graph that is to be traced.
6. The cursor is moved to points on the graph by pressing the right or left arrow keys.
7. With the graph of $y=2x^3-8x$ displayed, move the cursor to a point on the graph whose displayed x-coordinate is 1.111111111. The y-coordinate should read -6.145404664.
8. The coordinates of the points are displayed as you move the cursor.

1.7.2 TRACE A SCATTER PLOT
The TI-85 does not have a built-in trace feature for scatter plots. The program TRACE85 can be used to trace a scatter plot.
1. Down load the program TRACE85 from the program disk to your TI-85 by using TI-GRAPH LINK. If another person has downloaded this program to their TI-85, you can transfer the program to your calculator with a TI unit-to-unit link cable.
2. The lists that are used to graph a scatter plot must be stored in xT and yT in order to execute this program. List L1 in the walk-through example for Appendix 1.5 represents the independent variable and should be copied to xT and L2 to yT. Type the following keystrokes to store the data in xT and yT: (From the Home Screen)
    [ALPHA] [7] [1] [STO▸] [X-VAR] [-] [ENTER]
    [ALPHA] [7] [2] [STO▸] [2ND] [ALPHA] [0] [ALPHA] [-]
    [ENTER] (Notice that the TI-85 is case sensitive.)
3. Turn off all functions on the Y= menu.
4. From the home screen, press [PRGM] [F1] to list the names of programs that are currently stored in your calculator. If TRACE is not visible press [MORE] as many times as necessary to see TRACE. Press the appropriate function key (F1-F5) to select this program.
5. You should see TRACE85 displayed on the screen. Press [ENTER] to execute this program.
6. A scatter plot should be displayed with a menu at the bottom of the screen.
7. Press [F1] or [F2] to move the cursor.
8. Press [F3] to display the coordinates of a point marked by the cursor.
9. Moving the cursor to the left-most point and pressing [F3], you should see: x=.8, y=.4.
10. Press [ENTER] to return to the graph with the trace menu. Press [F4] to select Jump. The Jump Option (F4) lets you move the cursor quickly to a different point. The Jump screen tells how many points are contained in the scatter plot and the current position of the cursor, and prompts you to enter the number of the point you want to jump to. For example, entering 3 would jump the cursor to the third point.
11. When asked "Jump to:" type 3 and press [ENTER].
12. Press [F3] and you should see: x=1.2, y=1.4.
13. When the TRACE85 menu is displayed press [F5] to exit

TI-85

TRACE85 and return to the GRAPH screen.

2 Range Settings By changing the values of the Range variables you can display the desired portion of the coordinate plane on the graph screen. You can also determine the values of the independent variable x which are selected to plot the points of a graph. When you set values of Xmin and Xmax, the change in x, called Δx, is defined by the formula $\Delta x = \dfrac{Xmax - Xmin}{126}$. The change in y, Δy is defined by $\Delta y = \dfrac{Ymax - Ymin}{62}$.

2.1 STANDARD RANGE
The Standard Range is:

Xmin=-10	Ymin=-10
Xmax=10	Ymax=10
Xscl=1	Yscl=1

One way to obtain these range values is by pressing **[GRAPH] [F3]** to select Zoom and **[F4]** to select Zoom Standard. The graph screen with these range settings will be displayed.

2.2 TRIGONOMETRIC WINDOW (Radian Mode)
The Trig Range has the following settings:

$$Xmin = -\frac{63\pi}{24} \qquad Ymin = -4$$

$$Xmax = \frac{63\pi}{24} \qquad Ymax = 4$$

$$Xscl = \frac{\pi}{2} \qquad Yscl = 1$$

For these settings $\Delta x = \dfrac{\dfrac{63\pi}{24} - -\dfrac{63\pi}{24}}{126}$

$$= \dfrac{\dfrac{126\pi}{24}}{126}$$

$$= \dfrac{\pi}{24}$$

The choice of Δx is "nice" for the special angles. To obtain the Trig Range press **[GRAPH] [F3]** to select Zoom, **[MORE]** and **[F3]** to select Zoom Trig. The graph screen with these range settings will be displayed.

2.3 SQUARE WINDOW
ZSQR defines a window so that the ratio (xMax-xMin):(yMax-yMin) is about 3:2. This window takes into consideration the size of the screen to reduce graph distortion. To change to a square window press **[GRAPH] [F3] [MORE] [F2]**. The window variables will be changed and the current graph will be displayed.

2.4 USER-DEFINED RANGE
1. Press **[GRAPH] [F2]** to display the RANGE Editor. The cursor should be on the xMin= line. Type in your choice of values for xMin and press **[ENTER]**.
2. Your cursor should now be at the xMax= line. Type in your choice of values for xMax and press **[ENTER]**.
3. Continue until all window variables have the desired values. You may display a graph with this new window by pressing **[F5]** if a function is activated on the y= menu.

You can return to the home screen by pressing **[EXIT]**.

2.5 FRIENDLY WINDOW (centered about 0)
A friendly window centered about 0 has
　　　Xmin=-Xmax

so $\dfrac{2xMax}{126} = \Delta x$

　　　or xMax=63Δx
For example when Δx=.1
　　　Xmax=6.3
　　　Xmin=-6.3
For this set of range values a function is evaluated and graphed for x=-6.3, x=-6.2, ..., x=6.2, x=6.3. For this setting the trace mode produces 127 y values of a function for the above values of x. Note: You do not enter a value for Δx directly, but when Xmin and Xmax are set, Δx is automatically set.
(2.6 does not apply to the TI-85).

2.7 TURNING AXES OFF
To turn the coordinate axes off, press **[GRAPH] [MORE] [F3]** move the cursor down to AxesOn, right to AxesOff, and press **[ENTER]**. Press **[F5]** to display the graph.

3.1 SET ZOOM FACTORS
The Zoom factors determine the magnification when zooming in. They are initially set at 4 for Xfact and Yfact. Both Zoom factors are to be set at 10 for projects in this book.
1.　Press **[GRAPH]**, **[F3]**, **[MORE]**, **[MORE]** and **[F1]** to display the values of Xfact and Yfact.
2.　The cursor should be on the value of Xfact. Type in 10 and press **[ENTER]**. With the cursor on the value of Yfact, type in 10 and press **[EXIT]**. Your Zoom factors are now set at 10.

3.2 ZOOM IN ON A POINT
1.　Select a function on the Y= menu.
2.　Press **[GRAPH] [F3]** to select Zoom and **[F2]** to select Zoom In. The graph of the selected function will appear and your calculator is now in the Zoom-in mode.
3.　Use the cursor movement keys to move the cursor to a point which will be the center of the next viewing rectangle. This point is usually a point on the graph or close to a point on the graph.
4.　Press **[ENTER]** to display a "magnified" portion of the graph in a new viewing rectangle.
5.　You are still in the Zoom-in mode. To Zoom in again move the cursor to a point which will be the center of the next viewing rectangle (if necessary) and press **[ENTER]**.
6.　You may continue to Zoom in until you have a desired graph or you are beyond the range of your calculator.
7.　If you get an "Error 20 GRAPH RANGE", press **[F5]** to return to the Home screen.
8.　Press **[GRAPH] [F3]** to display the zoom menu. Press **[F5]** for Zoom Previous, which will return to the range settings prior to the last Zoom-in.

4 TI-Graph Link for the PC (Windows) If you have a DOS version of the Graph Link program, you will find instructions in your link manual.

4.1 PRINT A SCREEN IMAGE

1.   Boot up the Link-85 program on your computer.
2.   Display the calculator screen that you want to print.
3.   Attach the PC link cable to your calculator.
4.   Select Link from your computer screen menu to display the
     LINK menu and select "Get LCD from TI-85".
5.   When a sub menu is displayed on the computer screen, choose
     Printer (small) or Printer (large).
6.   When the Receive box is displayed, select **Receive**.
7.   When the LCD save box is displayed you should see the
     calculator screen display:
          Save the LCD image to:
          PRINTER
     At this time select ok.
8.   When the Print box is displayed, you may select the number
     of copies if you want more than one. Select **OK** to print
     out your screen display.
(Use the HELP option for more information).

4.2 LOAD A TI-85 PROGRAM FROM THE COMPUTER
1.   Boot up the Link-85 program on your computer.
2.   Attach the PC link cable to your calculator.
3.   Press **[2nd] [X-VAR]** on your calculator.
4.   Move the cursor to RECEIVE and press **[F2]**. You should see
     "Waiting".
5.   Select **[LINK]** and then **[SEND]** from your computer screen
     menu.
6.   The send box should be displayed. Select the appropriate
     drive and directory.
7.   Select **OK** to display a list of TI-85 file names on your
     disk. Select a file name. (press **CTRL** and select a file
     name if more than one file is to be sent).
8.   You should see the list of file names when the send box is
     displayed on your computer screen.
9.   At this time, select **OK** to transmit these files.
10. If a file is already stored in your calculator, you will
     have the option to:
          RENAM
          OVERW
          SKIP
          EXIT
11. Press the function key directly below your choice. If you
     choose "1. Rename", you will be prompted for a new name on
     your calculator screen.

4.3 RUN A PROGRAM
1.   The program must be stored in your calculator.   (See
     Appendix 4.2).
2.   Press **[PRGM]** and **[F1]** to display the names of all programs
     stored in your calculator.
3.   If you don't see the name of the program, press **[MORE]**
     until you do. If the name does not appear, you need to
     store the program in your calculator. (See Appendix 4.2).
4.   Press the function key directly below the program name.
5.   When the name is displayed on the home screen, press
     [ENTER] to run the program.
6.   If you get an error message you should press **[F5]** to quit.
7.   You can stop a program by pressing **[ON]**. When you get the
     BREAK message, press **[F5]** to quit.

5 Tables The TABLE option is not available as a built-in feature
on the TI-85. There are two programs TAB and TABLE85 which will

be used to provide a TABLE option.

5.1 TAB
The purpose of this program is to evaluate a function for values of the independent variable that are entered by the user.
Walk-through example: Evaluate the function $f(x)=3x^2-4$.

1. Download the program TAB from the program disk on your TI-85 by using the TI-GRAPH LINK. If another person has this program stored in their TI-85, you can transfer the program to your calculator with a TI unit-to-unit link cable.

2. Press **[GRAPH] [F1]** to display the y= edit screen.

3. The function must be stored at y1= to be compatible with this program. Place the cursor at y1= and press **[CLEAR]** to clear this function storage position.

4. Enter the following keystrokes: **[3] [X-VAR] [x²] [-] [4]** and press **[ENTER]**. Next press **[EXIT] [EXIT]** to return to the home screen.

5. Press **[PRGM] [F1]** to display the names of programs that are currently stored in your calculator. If TAB is not visible, press **[MORE]** as many times as necessary to see TAB.

6. Press the function key directly below TAB and then **[ENTER]** to execute the program.

7. When you see the prompt: ENTER THE NUM..., type in 4 and press **[ENTER]**.

8. At the prompt ENTER x, enter one of the numbers 5, -3, 1, 2 and press **[ENTER]** after each one. Notice that the number of values of the independent variable correspond to the four at the first prompt.

9. Press **[ENTER]** and your table should look like:
```
[[ 5 71]
 [-3 23]
 [ 1 -1]
 [ 2  8]]
```

5.2 TABLE-85
The purpose of this program is to produce a table of function values based on a smallest and largest value of the independent variable and a given increment. This program will clear any function that is stored at y1=.
Walk-through example: Evaluate the function $f(x)=3x^2-4$ for x=3, 3.5, 4, ... 5.5, 6.

1. Download the programs Table85 and VIEWS85 from the program disk to your TI-85 by using a TI-GRAPH LINK (see Appendix 4.2). The program Table85 cannot run without the sub-program VIEWS85.

2. Press **[PRGM] [F1]** to display the names of programs stored in your calculator. If Table is not visible, press **[MORE]** until it is.

3. Press the function key directly below Table and you should see "Table85" displayed. Press **[ENTER]** to execute the program.

4. Type 3 and press **[ENTER]** at TblMin=.

5. Type 6 and press **[ENTER]** at TblMax=.

6. Type .5 and press **[ENTER]** at ΔTbl=.

7. When y1= is displayed, type in the following keystrokes: **[3] [X-VAR] [x²] [-] [4] [ENTER]**.

8. You should see a table whose first column is 3, 3.5, 4, 4.5, 5, 5.5 and whose second column is the corresponding function values correct to four decimal places. The first value is 23.0000, the function value corresponding to 4.5

is 56.7500.

9.   Press **[F1]** to display the remaining pairs in the table.
10. If you have a long table, the Jump feature will let you jump to any position in the table.
11. Press **[F5]** (Exit) and you will be given a choice to reset the table **[F1]** (yes) or exit from the program **[F5]** (no).

6.1.1 STORE DATA IN TWO LISTS.

Walk-through example:
Store {.8, 1, 1.2, 1.5, 1.9} in list L1 and {.4, 1.1, 1.4, 2.3, 4} in list L2.

1.   Press **[STAT] [F2]** to choose EDIT from the STAT menu.
2.   At the bottom of the screen are the names of lists which are currently in use. By pressing MORE you can display all list names. The following steps assume that L1 and L2 are not current list names. If they are, go to Appendix 6.1.2. The cursor is currently at the top of the screen at a position for entering the list name for the xlist. Press **[7] [ALPHA] [1]** to name the x-list L1 and press **[ENTER]**. Press **[7] [ALPHA] [2]** to name the y-list L2 and press **[ENTER]**.
3.   Enter .8 for x1 and press **[ENTER]**. A "1" is automatically stored in y1. Type .4 over the "1" and press **[ENTER]**.
4.   Continue entering x-y pairs until all elements of the two lists are entered. Press **[EXIT]** two times to leave the stat mode.
5.   Press **[2nd] [-] [F3]** to display the list names currently in use. If you don't see L1 or L2, press **[MORE]** until you do.
6.   Press the appropriate function key (the one directly below the list name) to display the contents of lists L1 and L2. Compare your results with the above sets.

6.1.2 CLEAR A PAIR OF LISTS

1.   Press **[STAT] [F2]** to display the names of lists currently in use. You may have to press **[MORE]** to see all of the list names.
2.   Select the list names for xlist and ylist which you want to clear. You can choose the names at these positions by pressing the function key directly below the desired list. Press **[ENTER]** until you see the listing of the contents of the lists. You should see CLRXY above **[F5]**.
3.   Press **[F5]** (CLRXY) to delete the contents of these two lists. The list names are still on the name menu, even though they contain no data. You are now ready to store data in these lists. See Appendix 6.1.1.

6.2 THE ALGEBRA OF LISTS

Walk-through example:
To store the numbers 1, 2, 3, 4, 5 in list LA:
1.   Press **[2nd] [-]** to display the LIST menu.
2.   Press **[F4]** to display the prompt Name=.
3.   Press **[7] [LOG]** to name the list LA and press **[ENTER]**.
4.   You are now ready to store the above numbers in this list. Press **[1] [ENTER]** to assign 1 to the first element e1.
5.   Press **[2] [ENTER]** to assign the value of 2 to the second element. Continue until the numbers 1, 2, 3, 4, 5 are stored in list LA.
6.   Press **[EXIT] [2nd] [-]** to again display the LIST menu. Press **[F4]** to display the prompt Name=.

TI-85

7.   Press **[7]** **[SIN]** **[ENTER]** to create the list LB. Press **[EXIT]**.

8.   To double each number in list LA and store the results in list LB, press **[2nd]** **[-]** **[F3]** to display the names of currently used lists. Type 2, press **MORE**, if necessary to see LA, then press the function key below LA to select LA, and press **[STO▸]**. Press the function key below LB to select LB and press **[ENTER]**. You should see {2 4 6 8 10}, which is the new contents of LB.

9.   The algebra of lists can be extended to more complicated expressions such as $\ln(5x^2+4x+7)$. To apply this function to the numbers in list LA and store the results in LB, complete the following steps:
       Press **[2nd]** **[-]** **[F3]** to display the names of currently used lists.
       Press **[LN]** **[(]** **[5]** **[function key below LA]** **[x²]** **[+]** **[4]** **[function key below LA]** **[+]** **[7]** **[)]** **[STO▸]** **[function key below LB]** **[ENTER]**.
       You should see a list of 5 numbers. The first one should be 2.77258872224. Move the cursor to the right to display the other members of LB.

6.3 EDIT AN EXISTING LIST

1.   Press **[2nd]** **[-]** **[F4]** to put your calculator in the list edit mode and display the names of lists currently in use. You may have to press **[MORE]** to see all of the list names.

2.   Use the appropriate function key to select a list to edit and press **[ENTER]**.

3.   The values of list entries should be displayed. The up and down arrows move the cursor to the next list element.

4.   To move to the last element in a list, place the cursor on e1= and press the up arrow.

5.   You may type over, insert new elements or delete elements. When you have completed all changes press **[EXIT]** to exit the list edit mode.

Curve Fitting (Regression) A set of data must be stored in two lists of equal length before you can use your calculator to fit a curve to the data.

7.1 STORE DATA PAIRS IN TWO LISTS

Walk-through example:
Store data points in two lists L1={.8,1,1.2,1.5,1.9} and L2={.4,1.1,1.4,2.3,4} where the numbers in list L1 represent the independent variable and those in L2 the dependent variable. Follow the instructions in Appendix 6.1.1.

7.2 LINEAR REGRESSION

1.   After storing the data from Appendix 7.1 in L1 and L2, press **[STAT]** **[F1]** to display the current selection for the xlist and ylist as well as the names of other lists which contain data. Select the lists that represent the independent variable xlist (L1) and dependent variable ylist (L2) by pressing the appropriate function keys (F1-F5) and pressing **[ENTER]** each time.

2.   The CALC menu should appear at the bottom of the screen. Press **[F2]** to select LINR (linear regression). For the walk through example you should see:
       a=-2.22245989305
       b=3.17379679144
       corr=.989715285943

TI-85

n=5

The parameters a and b represent the y-intercept and slope of the line whose equation is y=a+bx. Corr is the correlation coefficient and is a measure of how well the line fits the data. The number of data points is n=5.

7.3 QUADRATIC REGRESSION

1. After storing the data from Appendix 7.1 in L1 and L2, press **[STAT] [F1]** to display the current selection for xlist and ylist as well as the names of other lists which contain data. Select the lists that represent the independent variable xlist (L1) and dependent variable ylist (L2) by pressing the appropriate function keys (F1-F5) and pressing **[ENTER]** each time.

2. From the CALC menu press **[MORE]** and **[F1]** to select P2REG, which is second degree (i.e. polynomial) regression. For the walk through example you should see:

    P2Reg
    n=5
    PregC={1.15772624556 .030600989993 -.269183794247}

    The three numbers in this set are values of a_2, a_1, and a_0 where $y=a_2x^2+a_1x+a_0$.

7.4 CUBIC REGRESSION

Proceed as in Appendix 7.3 and select P3REG. This option displays the values of a_3, a_2, a_1, and a_0 where $y=a_3x^3+a_2x^2+a_1x+a_0$ is the "best" cubic equation to fit your data.

7.5 QUARTIC REGRESSION

Proceed as in Appendix 7.3 and select P4REG. This option displays the values of a_4, a_3, a_2, a_1, and a_0 where $y=a_4x^4+a_3x^3+a_2x^2+a_1x+a_0$ is the "best" quartic equation to fit your data.

7.6 LOGARITHM REGRESSION

Proceed as in Appendix 7.3 and select LNR. This option displays a, b, corr, and n, where a and b are values of the parameters in y=a+blnx for x>0.

7.7 EXPONENTIAL REGRESSION

Proceed as in Appendix 7.3 and select EXPR. This option displays values of a, b, corr, and n, where a and b are values of the parameters in $y=ab^x$.

7.8 POWER REGRESSION

Proceed as in Appendix 7.3 and select PWRR. This option displays values of a, b, corr and n where a and b are values of the parameters in $y=ax^b$, x>0, y>0.

7.9 STORE A REGRESSION EQUATION ON THE Y= EDIT MENU

Walk-through example: Store the linear regression equation for the data set in Appendix 7.1.

1. Follow the curve-fitting process of Appendix 7.2.

2. Press **[GRAPH] [F1]** to display the y= edit screen and move the cursor to the right of y1= or to the right of any y= position which has no function currently stored.

3. Press **[STAT] [F5] [MORE] [MORE] [F2]**. You should see RegEq at the y= position marked by the cursor. Appendix 7.10 explains how to overlay the graph of this equation on the scatter plot.

TI-85

7.10 OVERLAY THE GRAPH OF A REGRESSION EQUATION ON A SCATTER PLOT

Walk-through example: Graph the regression line which was stored in Appendix 7.9.

1.   Follow the instructions in Appendix 7.9 to store the linear regression equation for the data in the walk through example.
2.   Set your range variables to fit your data.
3.   Press **[STAT] [F3] [F2] [CLEAR]**. You should see the graph of the regression equation overlaid on the scatter plot. (The points of the scatter plot will probably be hard to see.) You can run TRACE85 to trace the points of the scatter plot (see Appendix 1.7.2). Your graph should look similar to the one in Figure 1.

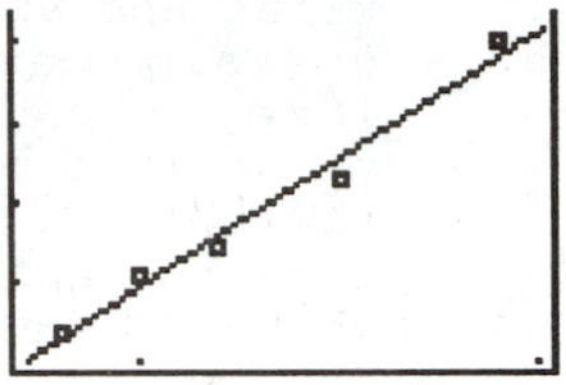

Figure 1

8.1 NDER (Numerical Derivative)

1.   Press **[2nd] [÷]** to display the CALC menu.
2.   Choose nDer by pressing **[F2]**.
3.   The syntax for the nDer command is: nDer (f(x),x,a). For example nDer $(3x^2+x,x,2)$ returns an approximation to f'(x) at x=2 where $f(x)=3x^2+x$. With "nDer(" displayed enter the following keystrokes: **[3] [x-VAR] [x²] [+] [x-VAR] [,] [x-VAR] [,] [2] [)] [ENTER]**. After completing the nDer command, you should see nDer $(3x^2+x,x,2)$ 13.

8.2 $\frac{dy}{dx}$ (Numerical Derivative)

Walk-through example: Graph $y=3x^2+x$ with a friendly window of xMin=-6.3, xMax=6.3 and follow the directions below to approximate the slope of the tangent line to this graph at x=1.

1.   Press **[GRAPH]**, **[MORE]** and **[F1]** to display the GRAPH MATH menu. (Graphs of activated functions on the Y= menu are displayed).
2.   Choose dy/dx by pressing **[F4]**.
3.   If more than one graph is displayed choose the desired one by using the up or down arrow **[▲]** or **[▼]**. (You need at least one activated function on the Y= menu).
4.   Use the right or left arrow **[▶]** or **[◀]** to trace to the point where x=1 and press **[ENTER]**.
5.   The numerical derivative of the function at x=1 should be dy/dx=7.

8.3 $\frac{dr}{d\theta}$

Walk-through example: Approximate $\frac{dr}{d\theta}$ at $\frac{\pi}{6}\approx.524$ for r=4sin(2θ).

1.   Follow the directions with specified window settings in Appendix 1.4 to graph r=4sin(2θ).
2.   With the graph displayed and only the GRAPH menu showing, press **[MORE] [F3]** to display the FORMAT menu.
3.   Position the cursor over PolarGC and press **[ENTER]**.
4.   Press **[F5]** to display the graph of r=4sin(2θ), press **[MORE] [F1] [F3]** to select $\frac{dr}{d\theta}$.
5.   Move the cursor to the point whose θ-coordinate is about

.524 and press **[ENTER]**. You should see dr/dθ=4.

8.4 FNINT (Numerical Integral)

Walk-through example: Approximate $\int_1^3 x^3\, dx$

1.   Press **[2nd]** **[÷]** to display the CALC menu and press **[F5]** to choose fnInt.
2.   The syntax for fnInt is fnInt(f(x),x,a,b). For example, fnInt(x^3,x,1,3) returns an approximation of the integral $\int_1^3 x^3\, dx$. After completing the fnInt command, press **[ENTER]** to display the result which is 20.

8.5 ∫f(x)dx (Numerical Integral from the graph of y=f(x))

Walk-through example: Approximate $\int_1^3 x^3\, dx$.

1.   Store y=x^3 on the y= menu and select this function for graphing.
2.   Set the Range variables as:
      xMin=-6.3   yMin=-5
      xMax= 6.3   yMax=30
      xScl= 1   yScl=5
3.   Press **[GRAPH]** **[MORE]** **[F1]** to display the GRAPH MATH menu. Press **[F5]** and you will see a graph of the function displayed.
4.   Use the right arrow key to move the cursor to a point whose x-coordinate is 1 and press **[ENTER]** (1 is the lower limit of integration).
5.   Use the right arrow key to move the cursor to a point whose x-coordinate is 3 and press **[ENTER]**. (This marks the upper limit of integration.) Press **[ENTER]** to approximate $\int_1^3 x^3\, dx$. The approximate value of 20 is displayed in the lower left portion of the screen.

8.6 DRAW A TANGENT LINE

Walk-through example: Draw a line tangent to the polar graph of r=4sin(2θ) at the point where θ=$\frac{\pi}{6}$≈.523.

1.   Follow the instructions in Appendix 1.4 to graph the above function. Set the trace to display polar coordinates. (See Appendix 1.7.1)
2.   Press **[GRAPH]** **[MORE]** **[F1]** to display the math menu and **[F5]** to select TANLN.
3.   Use the right arrow key to move the cursor to the point whose θ coordinate=$\frac{\pi}{6}$≈.524 and press **[ENTER]**.

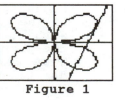

Figure 1

4.   Your graph should look like the one in Figure 1.

9.1 THE SEQUENCE COMMAND

The seq(command is used to evaluate selected terms of a sequence or store selected terms of a sequence in a list.
Walk-through example: Display the 3rd, 5th, 7th, 9th, and 11th terms of the sequence defined by $a_n=n^3-2$.

TI-85

1. Press **[2nd] [x]** to display the MATH menu.
2. Press **[F5]** (misc) and **[F3]** (seq) to display seq(.
3. Enter the following keystrokes: **[x-VAR] [∧] [3] [-] [2]
    [,] [x-VAR] [,] [3] [,] [1] [1] [,] [2] [)]**. You should
    see seq(x∧3-2,x,3,11,2). The first position is the
    sequence formula, the second the variable, the number of
    the first term to be calculated, the number of the last
    term to be calculated and the 2 is the increment. This
    instruction displays the values of a_3, a_5, a_7, a_9, and a_{11}.
4. Press **[ENTER]** and you should see {25 123 341 727 1329}

9.2 THE SUM COMMAND

Walk-through example: Calculate $\displaystyle\sum_{n=1}^{5} \frac{1}{n} = 1 + \frac{1}{2} + \frac{1}{3} + \frac{1}{4} + \frac{1}{5}$.

1. Press **[2nd] [x]** to display the MATH menu.
2. Press **[F5]** (misc) **[F1]** to display sum.
3. Press **[F3]** (seq) and "sum seq(" with a blinking cursor
    should be displayed on the home screen. This is a prompt
    to enter the parameters of the seq command. Enter the
    following keystrokes: **[1] [÷] [x-VAR] [,] [x-VAR] [,] [1]
    [,] [5] [,] [1] [)] [ENTER]**.
4. The sum of 2.28333333333 should be displayed.

9.3 GRAPH AN EXPLICIT SEQUENCE WITH RECSEQ

The program RECSEQ contains some features that are not
available in parametric graphing.

Walk-through example: Graph the first 20 terms of the sequence
$a_n = \dfrac{n-2}{n}$.

1. Down load the programs RECSEQ, VIEWS85, and TRACE85 from
    the program disk to your TI-85 by using a TI-GRAPH LINK.
    The RECSEQ program cannot run without the two subprograms
    VIEWS85 and TRACE85. If another person has these
    programs stored in their calculator, you can transfer
    them to your calculator with a TI unit-to-unit cable.
2. Press **[PRGM] [F1]** to display the names of programs that
    are currently stored in your calculator. If RECSEQ is
    not visible, press **[MORE]** until you see RECSEQ. Press
    the function key directly below RECSEQ and you should see
    RECSEQ. Press **[ENTER]** to execute this program.
3. With the cursor at UN= enter the following keystrokes to
    define the sequence: **[(] [ALPHA] [9] [-] [2] [)] [÷]
    [ALPHA] [9]**.
    You should see UN=(N-2)/N. It is mandatory that you use
    N for the variable in this program rather than n or any
    other character.
4. Press **[ENTER]**, type the value of the first term of the
    sequence which is -1, and press **[ENTER]**.
5. At the prompt "Start at?" type 1, since we want to begin
    with the first term. Press **[ENTER]** and you should see
    the prompt "Stop at?"
6. Type 20 for the number of the last term, since we want
    the first 20 terms. Press **[ENTER]**.
7. The screen should now read:
    　　1: Start over

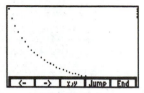

```
                          2: Graph
                          3: Table
                          4: Quit
                          Choice?
```
        Press **[2] [ENTER]** to graph the first
        20 terms of the sequence.

Figure 1

8.   Your screen should look like the one
        in Figure 1.
9.   Press **[F2] [F2]** to move the cursor to
        the third point on the graph and press **[F3]** to display
        the number of the term and the value of that term. You
        should see:
```
                          x=3
                          y=.3333333333323
                          Press "Enter".
```
10.   Press **[ENTER]** to return to the graph. Press **[F4]** to
        display:
```
                          # of items: 20
                          current:      3
                          Jump to:
```
        At the prompt type 9 and press **[ENTER]** to jump to the 9th
        term. Press **[F3]** and you should see:
```
                          x=9
                          y=.777777777778
```
11.   Press **[ENTER]** to return to the graph menu. Press
        [F5] (End) to return to the options menu. From this
        screen you can generate a table or a graph for a new
        sequence (option 1), review the existing graph or table
        (option 2 or 3), or exit the program (option 4).

9.4 GRAPH SELECTED TERMS OF A RECURSIVE SEQUENCE

Walk-through example: Graph a_{10}, a_{11}, a_{12}, ... a_{50} for the recursive

sequence $a_n = \dfrac{a_{n-1}+1}{n}$, $a_1=2$

1.   Follow steps 1 and 2 of Appendix 9.3 to start the program
        RECSEQ.
2.   With the cursor at UN= enter the following keystrokes to
        define the sequence: **[(] [ALPHA] [1] [ALPHA] [7] [+] [1]
        [)] [÷] [ALPHA] [9]**
        You should see UN=(UL+1)/N
        It is mandatory for this program that you use the symbol
        UL for a_{n-1} and the symbol N to represent the independent
        variable.
3.   Press **[ENTER] [2] [ENTER]** to enter the
        value of the first term.
4.   At the prompt "Start at?", enter 10
        and press **[ENTER]** (Check the example).
5.   At the prompt "Stop at?", enter 50 for
        the number of the last term and press
        [ENTER].
6.   After a moment you should see the
        options menu:

Figure 2

```
                          1: Start over
                          2: Graph
                          3: Table
                          4: Quit
                          Choice?
```
        Press **[2] [ENTER]** to choose the graph option.
7.   Your screen should be like the one in Figure 2.
8.   Press **[F2]** three times to move the cursor to the fourth
        point on the graph. Press **[F3]** to check your result.

You should see:
        x=13
        y=.083981783873
This is interpreted as a_{13}=.083981783873

9.   Press **[ENTER]** to return to the graph. Press **[F4]** to display:
        # of items: 41
        Current: 4
        Jump to:
      At this prompt, type 17 and press **[ENTER]** to "Jump" to the 17th point which is a_{26}. Press **[F3]** and you should see:
        x=26
        y=.040067022668

10.   Press **[ENTER]** **[F5]** (Exit) to return to the options menu. From this screen you can generate a table or a graph for a new sequence (option 1). Review the existing graph or table (options 2 or 3) or exit the program (option 4).

9.5 DISPLAY THE VALUES OF THE TERMS OF A SEQUENCE WITH RECSEQ
Walk-through example: calculate and display the value of the first 20 terms of the sequence $a_n = \dfrac{n-2}{n}$.

1.   Follow steps 1-6 of Appendix 9.3 and press **[ENTER]**. You should see the following display:
        1: Start over
        2: Graph
        3: Table
        4: Quit

2.   Press **[3]** **[ENTER]** to display two columns of numbers headed by IND VAR and DEP VAR. The numbers in the first column represent the number of the term. The corresponding number in the second column is the value of that term. For this example a_5=.6000 and a_6=.6667. (The value of each term is rounded to 4 decimal places.)

3.   Press **[F1]** three times to scroll through this table. The bottom row of numbers should be 9 |.7778 or a_9=.7778.

4.   Press **[F3]**, enter 18 and press **[ENTER]** to "Jump" through this table. The last row should be 20 |.9000 which is interpreted as a_{20}=.9000.

5.   Press **[F5]** (Exit) to return to the options menu. From this screen you can generate a table and/or graph for a new sequence, review the existing table or graph, or exit the program by pressing **[4]** **[ENTER]**.

9.6 DISPLAY THE VALUES OF THE TERMS OF A RECURSIVE SEQUENCE
Walk-through example: Display the value of a term of a recursive sequence $a_n = \dfrac{a_{n-1}+1}{n}$, a_1=2.

1.   Follow steps 1-5 of Appendix 9.4. You should see the options menu:
        1: Start over
        2: Graph
        3: Table
        4: Quit
        Choice?
      Press **[3]** **[ENTER]** to choose the Table option.

2.   You should see two columns of numbers headed by IND VAR and DEP VAR. The numbers in the first column represent the number of the term. The corresponding number in the

second column is the value of that term.   For this
example a_{10}=.1127 and a_{15}=.0718.

3.   Press **[F1]** three times to scroll through the table.   The
bottom row of numbers should be 18 |.0590 or a_{18}=.0590.

4.   Press **[F3]**, enter 35 and press **[ENTER]** to "Jump" through
the table so that the top row is 35 |.0294 or a_{35}=.0294.

5.   Press **[F5]** (Exit) to return to the options menu.   From
this screen you can generate a table or a graph for a new
sequence (option 1), review the existing graph or table
(options 2 or 3), or exit the program (option 4).

9.7 PARAMETRIC GRAPHING OF A SEQUENCE

Walk-through example: Graph 20 terms of the sequence $a_n = \dfrac{n-2}{n}$.

1.   Press **[2nd] [MORE]** to display the MODE screen.
2.   Move the cursor down to Func and over to Param and press
[ENTER] to change to the parametric graphing mode.
3.   Press **[GRAPH] [F1]** to display the E(t)= edit screen.
Place the cursor at xt1= and press **[F1] [ENTER]**.
4.   Enter the following key strokes: **[(] [F1] [-] [2] [)] [÷]**
[F1].   The parametric form of the sequence is stored for
graphing.
5.   The "=" should be highlighted for xt1 and yt1.   If not,
place the cursor on any symbol at xt1= or yt1= and press
[F5].
6.   Press **[EXIT] [F2]** to display the RANGE variables edit
screen.   Enter each value below and press **[ENTER]** until
the range variable have the following values.

        tMin=1   xMin=1   yMin=-2
        tMax=20   xMax=20   yMax=2
        tStep=1   xScl=1   yScl=1

7.   Press **[MORE] [F3]** to display the FORMAT screen.   Move the
cursor down to Drawline, over to DrawDot, and press
[ENTER] to change the graphing mode to just plotting
points and not connecting them.
8.   Press **[F5]** to display a graph of the sequence.   Your
graph should look line the one in Figure 3.

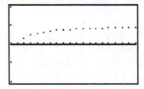

Figure 3

9.   Press **[F4]** to activate the TRACE and press the right
arrow key until t=7.   You should see:
        t=7
        x=7   y=.71428571429
The value of x is the same as t which is the number of
the term, 7 in this case.   The y-value is the value of
the seventh term, a_7.

10.1 STORE A PICTURE

Walk-through example: To store the graph of $y=x^2$ as a picture:
1.   Store $y=x^2$ at y1=, set a standard window (ZStd), and
display its graph. (See Appendix 1.1)   If other functions
are graphed, go back to the y= editor and deselect them.
2.   Press **[GRAPH]** and when the graph is displayed press
[MORE] twice.   You should see STPIC above the F2 key.
3.   Press **[F2]** and you should see the prompt "Name=".   You

may type in a name or select one of the PIC names from
the menu (if there are any).
4.   Press [,] [)] [COS] [ALPHA] [1] to name the picture PIC1
      and then press [ENTER].

10.2 RECALL A PICTURE
Walk-through example: To recall the picture that was stored in
Appendix 10.1:
1.   Press [GRAPH] [F1] and deselect any function that is
      activated for graphing.
2.   Press [STAT] [F3] [F5] to clear any drawings from the
      screen.
3.   Press [GRAPH] to display the GRAPH menu and press [MORE]
      twice. You should see STPIC and RCPIC on the menu.
4.   Press [F3] (RCPIC) and the prompt "Name=" should be
      displayed.
5.   PIC1 should appear on the list of PIC names. Press the
      function key directly under PIC1 and press [ENTER] and
      the graph of $y=x^2$ should be displayed.

10.3 OVERLAY A PICTURE ON A GRAPH
Walk-through example: Overlay the picture which was stored in
Appendix 10.1 over the graph of $y=(.3)^x$.
1.   Follow Appendix 10.1 to store the graph of $y=x^2$ with a
      standard window as a picture with name PIC1.
2.   Store $y=(.3)^x$ on the y= editor. Display its graph with
      a standard window.   (Deselect any other functions).

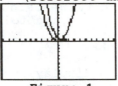

Figure 1

3.   Recall the picture named PIC1 (see Appendix 10.2).
4.   Your graph should look like the one in Figure 1.
5.   The window settings for the current graph must be the
      same as the picture in order for the graph to be
      meaningful.

11.1 APPROXIMATE A SOLUTION TO AN EQUATION
Walk-through example: Solve $\cos x = x-1$, $-10 \le x \le 10$.
1.   Write the equation as $\cos x - x + 1 = 0$.
2.   Press [GRAPH] [F1] to display the y(x)= edit screen.
3.   Store $y = \cos x - x + 1$ at a y= position and deselect any other
      functions which may be stored.
4.   Press [EXIT] [F3] [F4] to set up a standard graphing
      window and display the graph.
5.   With the GRAPH menu displayed, press [MORE] [F1].
6.   Press [F1] (LOWER), position the cursor to the point
      whose x-coordinate is 0 and press [ENTER].
7.   Press [F2] (UPPER), move the cursor to a point whose
      x-coordinate is about 2 and press [ENTER] By setting
      LOWER and UPPER an interval which contains the desired
      x-intercept is defined..
8.   Press [F3] (Root) [ENTER] to approximate the y-intercept
      of the graph which corresponds to the solution of the
      equation $\cos x = x-1$, $-10 \le x \le 10$.   The   solution,
      x=1.2834287417 should be displayed. (You can select the
      number of digits to be displayed by selecting Float on
      the MODE menu.)

11.2 APPROXIMATE A SOLUTION TO A SYSTEM OF TWO EQUATIONS
Walk-through example: solve y=cosx and y=x-1 for -10≤x≤10.
1.   Write the equation cosx=x-1 and follow B11.1.

1.1 FUNCTION GRAPHING

Walk-through example: Graph the function defined by $f(x)=2x^3-8x$.

1. Press **[MODE]** to check the current graphing mode. The cursor should be on the Graph line. Press the right edge of the cursor pad to display the menu of graphing modes. Move the cursor to option 1 to highlight the FUNCTION graphing mode and press **[ENTER]** twice to save this mode selection.

2. Press **[APPS]** and press **[2]** to display the Y= Editor. Move the cursor to y1 and press **[CLEAR]** to delete the function stored at y1= or move the cursor to a vacant storage position.

3. Enter the following keystrokes: **[2]** **[X]** **[∧]** **[3]** **[-]** **[8]** **[X]**. You should see y1(x)=2x∧3-8x on the edit line at the bottom of the screen. If so, press **[ENTER]** to store the function at y1=. You should have a check (✓) by the y1 which indicates that this function is selected for graphing. If not, move the cursor to the y1 line and press **[F4]** which acts like an on-off switch for selecting or deselecting a function for graphing.

4. Move the cursor to any other lines which have a check [✓] and press **[F4]** to turn them off.

5. Define the viewing rectangle by pressing **[F2]** **[6]** to select a standard window (see Appendix 2.1).

6. You should see a graph of $y=2x^3-8x$ for the portion of the coordinate plane defined by the standard window setting. Your graph should look like the one in Figure 1.

7. You can graph more than one function on the same screen by selecting several functions on the Y= Editor.

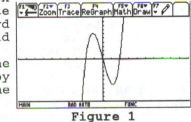

Figure 1

1.2 PIECEWISE FUNCTIONS

Walk-through example: $y=\begin{cases} 3x-2, & x<2 \\ x^3-2, & x\geq 2 \end{cases}$

1. Press **[APPS]** **[2]** to display the y= Editor. Place the cursor at y1=, press **[CLEAR]**, and enter the following keystrokes: **[W]** **[H]** **[E]** **[N]** **[(]** **[X]** **[2ND]** **[5]** **[8]** **[2]** **[2]** **[,]** **[3]** **[X]** **[-]** **[2]** **[,]** **[X]** **[∧]** **[3]** **[-]** **[2]** **[)]** **[ENTER]**. You should see
y1= $\begin{cases} 3x-2, & x<2 \\ x^3-2, & \text{else} \end{cases}$ on the y= Editor screen.

2. Define the viewing window by pressing **[F2]** **[6]** for the standard window. In some cases you may need to adjust your window variables to produce a better graph (see Appendix 2.4).

3. A piecewise function graph sometimes looks more realistic if the points are not connected. To set the dot mode rather than the connected mode, press **[APPS]** **[2]** to return to the y= Editor, and move the cursor to the function that you want to change to dot.

4. Press **[F6]** for the style menu and press **[2]** for Dot. Press **[APPS]** **[4]** to graph the function. Your graph should look like the one in Figure 2.

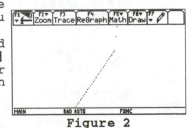

Figure 2

1.3 PARAMETRIC GRAPHING

Walk-through example: Graph the function $y=2x^3-8x$ parametrically.

1. Change the form of this equation by letting $x=t$ and $y=2t^3-8t$.

2. Change your calculator to the parametric mode by pressing **[MODE]** and using the cursor pad to move the cursor to the right to display the graphing mode menu. Press **[2]** to choose the PARAMETRIC GRAPHING MODE, and **[ENTER]** to save this setting.

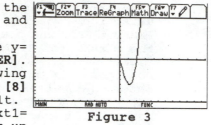

Figure 3

3. Press **[APPS]** **[2]** to display the y= Editor. At xt1= press **[T]** **[ENTER]**.

4. At yt1= enter the following keystrokes: **[2]** **[T]** **[∧]** **[3]** **[-]** **[8]** **[T]** **[ENTER]** and check your result.

5. Your should see a check mark by xt1= and yt1=, if not, move the cursor up to highlight xt1 and press **[F4]**. Then highlight yt1 and press **[F4]**.

6. Press **[APPS]** **[3]** to display the Window Editor.

7. The cursor should be on tmin, type 0 and press **[ENTER]**. Continue typing in values and pressing **[ENTER]** to complete each entry with the values below:

    tmin=0   xmin=-10   ymin=-10
    tmax=6   xmax=10   ymax=10
    tstep=.1   xscl=1   yscl=1

8. Press **[APPS]** **[4]** to graph the function and compare your results with the graph in Figure 3. In the parametric mode you can control the number of points that are graphed. For the above settings, a point is plotted for t=0, t=.1, t=.2 up through t=6 for a total of 61 points.

1.4 POLAR GRAPHING

Walk-through example: Graph $r=4\sin(2\theta)$

1. Change your calculator to the polar graphing mode by pressing **[MODE]** then moving the cursor to the right to display the graphing mode menu. Press **[3]** for the POLAR mode.

2. Move the cursor down to the RADIAN/DEGREE line and to the right to display the angle measure mode and press **[1]** to select RADIAN.

3. Press **[ENTER]** to save these changes.

4. Press **[APPS]** **[2]** to display the y= Editor. At r1= enter the following keystrokes: **[4]** **[SIN]** **[2]** **[θ]** **[)]** **[ENTER]**. You should see √r1=4·sin(2·θ).

5. Press **[F2]** **[6]** to set the standard window for polar graphing and display the graph of $r=4\sin(2\theta)$.

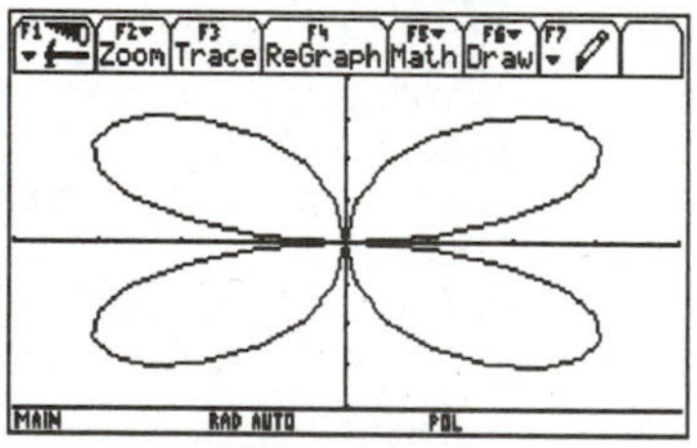

Figure 4

6. This graph can be improved by changing the window settings. Press **[APPS]** **[3]** to display the window editor. Move the cursor and make the following changes:

    xmin=-4   ymin=-4
    xmax=4   ymax=4

Press **[APPS]** **[4]** to display your graph which should look like the one in Figure 4.

1.5 SCATTER PLOTS

The scatter plot feature is used to plot points from two lists of

TI-92

the same length.

Walk-through example: Graph a scatter plot of lists L1= {.8, 1, 1.2, 1.5, 1.9} and L2={.4, 1.1, 1.4, 2.3, 4}. Where the numbers in L1 represent the independent variable and those in L2 represent the dependent variable.

1. Press **[APPS] [1] [CLEAR]** to display the home screen and clear the command line.

2. List L1 is created by using the **[STO▸]** command.

3. Create list L1 by entering the following keystrokes: **[CLEAR] [2ND] [(] [.] [8] [,] [1] [,] [1] [.] [2] [,] [1] [.] [5] [,] [1] [.] [9] [2ND] [)] [STO▸] [L] [1] [ENTER]**. You should see {.8,1,1.2,1.5,1.9}→l1 followed by {.8,1,1.2,1.5,1.9}.

4. Create list L2 by storing the second set of numbers in L2, using Step 3 as an example. When you are finished check the numbers in this list with the display on the home screen.

5. A scatter plot will now be graphed by using lists L1 and L2. Press **[MODE]** and your cursor should be on the Graph line. Move the cursor to the right to display the types of graphing modes. Move the cursor to Function and press **[ENTER]**. Press **[ENTER]** again to save this choice.

6. Press **[APPS] [2]** to display the y= Editor. Move the cursor upwards until you see Plot1: highlighted.

7. Press **[F3]** to display the dialog box for plots. The cursor should be on the Plot type line. If Scatter is highlighted, move the cursor down to the Mark line. If not, move the cursor to the right, place the cursor on Scatter and press **[ENTER]**, and then move the cursor down to the Mark line.

8. Move the cursor to the right to display the selection menu for the type of symbol to mark the points on the scatter plot. Place the cursor on choice 1:Box and press **[ENTER]**.

9. Move the cursor down to the x line and press **[L] [1] [ENTER]** to enter list L1 for the x values (independent variable).

10. Move the cursor down to the y line and press **[L] [2] [ENTER]** to enter L2 for the y values (dependent variable).

11. NO should be highlighted for the Use Freq and Categories. If not, move the cursor down to that line, to the right, put the cursor on 1:NO, and press **[ENTER]**.

12. Press **[ENTER]** to save the dialog box selections.

13. You should be at the y= Editor screen and there should be a check mark by Plot:1. If not, press **[F4]** to check Plot:1 for graphing.

14. If there are other functions with check marks, move the cursor to these functions and press **[F4]** to deactivate them.

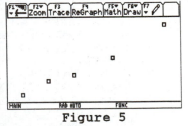

Figure 5

15. Press **[F2]** to display the zoom options and press **[9]** to choose window settings appropriate for the data in lists L1 and L2. You should see a scatter plot that looks like the one in Figure 5.

1.6 GRAPHS AND SCATTER PLOTS

Sometimes you may want to compare the graph of a function on the y= Editor with a scatter plot.

1. Press **[APPS] [2]** to display the y= Editor and store y=3x-2 at one of the function storage positions by entering **[3]**

[X] [-] [2] [ENTER]. This function should now have a check mark for graphing.

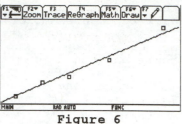

Figure 6

2.   Follow the steps in Appendix 1.5 to set up and graph a scatter plot. Your graph of the line overlaid on the scatter plot should look like the one in Figure 6.

1.7 TRACE
The trace feature can be used in FUNCTION, PARAMETRIC, POLAR, SEQUENCE and SCATTER PLOT graphing modes.
1.   Display the graph which you want to trace (see Appendices 1.1-1.6, 9.3, or 9.4).
2.   Press **[F3]** to activate the trace and notice the symbol in the upper right hand corner. P1 indicates that Plot 1 is graphed. A "2" indicates that the function stored at y2 is graphed.
3.   If more than one graph or scatter plot is displayed, press up or down on the cursor pad to select the function that you want to trace.
4.   Move the cursor to the right or the left to trace along a graph.
5.   The coordinates of a point on the graph marked by the cursor are displayed. The display depends on whether you are in the FUNCTION, PARAMETRIC, POLAR, or SEQUENCE mode.
6.   When you are tracing, you can display rectangular or polar coordinates. From the GRAPH screen, press **[F1]** **[9]** to display the GRAPH FORMAT screen. With the cursor on the coordinates line, move it to the right. You can choose 1 to display rectangular or 2 or display polar coordinates. Press **[ENTER]** to return to the graph.

2 Window Variables
By changing the values of the window variables, you can display the desired portion of the coordinate plane on the graph screen. You can also determine the values of the independent variable, x, which are chosen to plot the points of a graph. When you set the values of xmin and xmax the change in x, called Δx, is automatically determined by the formula $\Delta x = \dfrac{xmax - xmin}{238}$. For a function y=f(x), the coordinates of the first point are (xmin, f(xmin)), the second are (xmin+Δx, f(xmin+Δx)), the third (xmin+2Δx, f(xmin+2Δx)) and the trend continues to the last point (xmin+238Δx, f(xmin+238Δx)) which is (xmax, f(xmax)).

2.1 STANDARD WINDOW
1.   The standard window for the FUNCTION mode is:
            xmin=-10   xmax=10   xscl=1   xres=2
            ymin=-10   ymax=10   yscl=1
      The distance between tick marks on the x-axis is xscl and on the y-axis yscl.
2.   The standard window for Parametric and Polar graphing is the same as the one for Function graphing except:
            Parametric   Polar
            t=0   θ=0
            t=2π   θ=2π
            tstep=π/24   θstep=π/24
3.   To obtain a standard window, press **[APPS]** **[3]** to display the Window Editor. Press **[F2]** to choose the zoom menu and

[6] to set the standard window. The standard graph viewing window will be displayed.

2.2 TRIGONOMETRIC WINDOW (Radian Mode)

The Trig window has the following settings:

xmin=-15.577063574   ymin=-4   xres=2
xmax= 15.577063574   ymax=4
xscl=1.5707963267949   yscl=.5

This window is centered about 0 and the xmin and xmax are chosen so that $\Delta x = \pi/24$. This is a "nice" window for trigonometry because the special angles are multiples of $\pi/24$, therefore the special angles are selected as x coordinates of points to be plotted. The xscl is chosen so that the tick marks are every $\pi/2$ units so the quadrants are marked with a tick mark.

To obtain a Trig Window press [APPS] [3] [F2] to display the zoom menu. Press [7] (ZoomTrig). The viewing rectangle with the above settings is displayed.

2.3 SQUARE WINDOW

ZoomSqr defines a window which adjusts the x and y window settings to match the dimensions of the screen so that graphs do not look distorted.

To change to a square window press [APPS] [3] [F2] to display the zoom menu and press [5] (ZoomSqr).

2.4 USER-DEFINED WINDOW

There are instances for which you will want to choose values for the window variables that are different than those in a zoom window.

1.   Press [APPS] [3] to display the Window Editor. The cursor should be on the xmin for the Function graphing mode.
2.   Enter your value for this variable and press [ENTER].
3.   Enter your choice for the value of the next variable and again press [ENTER].
4.   Continue in this manner until all values of the window variables are entered. To display the graph with your new window settings press [APPS] [4].
5.   The distance between tick marks on the x-axis is xscl and on the y-axis is yscl. If your x- or y-axis looks thick, you may be assigning too many tick marks. For example, if xmin=-100, xmax=100, and xscl=1 there would be 200 tick marks on the x-axis.
       The variable xres controls how many points are actually graphed. For xres=1 all 239 points are plotted. For xres=2, every other point is graphed and for xres=10 every tenth point is graphed.

2.5 FRIENDLY WINDOW (centered about 0)

A window centered about 0 has xmax=-xmin so $\Delta x = \dfrac{2\,(xmax)}{238}$ or

$xmax = \dfrac{238}{2}\Delta x = 119\Delta x$. To set up a window with $\Delta x = .1$ let xmin=-11.9 and xmax=11.9. This window produces a graph whose chosen x-coordinates are all numbers between -11.9 and 11.9 which differ by .1. For a friendly window with $\Delta x = .01$ let xmin=-1.19 and xmax=1.19.

2.6 STAT WINDOW

To obtain this window press [APPS] [3] [F2] to display the zoom menu. Press [9] (Zoom Data). The viewing rectangle with window settings that include all the data in the scatter plot displayed.

TI-92

2.7 TURNING AXES OFF
To turn the coordinate axes off, press **[APPS]** **[4]** to display the graph and choose **[F1]** **[9]** to view the graphing format. Move the cursor down to Axes, right to display the drop-down box, and press **[1]** **[ENTER]** to turn the axes off.

3.1 SET ZOOM FACTORS
The Zoom Factors determine the magnification when using the ZoomIn feature. The Zoom factors are initially set at 4 for xFact and yFact. The Zoom factors for projects in this book are to be set at 10 for both xFact and yFact.
1.   Press **[APPS]** **[3]** **[F2]** to display the zoom menu and press **[C]** (Set Factors) to display the ZOOM FACTORS dialog box.
2.   The cursor should be on xFact. Type **[1]** **[0]** and move the cursor down to yFact.
3.   Type **[1]** **[0]** for yFact and press **[ENTER]**. Since our 2-dimensional graphs are not affected by zFact, press enter again to save the changes. This will return you to the Graph screen.

3.2 ZOOM IN ON A POINT
1.   Select a function at the y= Editor and display its graph.
2.   Change the Zoom Factors to 10 if you haven't done so (see Appendix 3.1).
3.   Press **[F2]** to display the zoom menu and press **[2]** to select ZoomIn.
4.   Move the cursor to a point on or close to the graph which is to be the center of the next viewing rectangle. Press **[ENTER]** to magnify the graph about its new center.
5.   Repeat steps 3 and 4 until you have a desired graph or you are beyond the range of your calculator.
6.   If you get a window variable domain error, you have probably zoomed once too often, so that your xmin and xmax are essentially the same. To remedy this situation, press **[ESC]** to return to the graph screen. Press **[F2]** and move the cursor down to **B:Memory** on the zoom menu. Move the cursor to the right and highlight the first option ZoomPrev. Press **[1]** to return to the Graph screen with the previous window settings.

4 TI-Graph Link for the PC (Windows) If you have a DOS version of the graph Link program, you will find instructions in your link manual.

4.1 PRINT A SCREEN IMAGE
1.   Boot up the Link-92 program on your computer.
2.   Display the calculator screen that you want to print.
3.   Attach the PC link cable to the I/O port on your calculator.
4.   Select **Link** and then **GET SCREEN** from your computer screen menu.
5.   At the Get screen box, select **GET SCREEN**.
6.   When a second Get screen box is displayed, you should see your calculator screen displayed on the computer. At this time select **PRINT**.
7.   When the Print dialog box is displayed, you can select the number of copies to be printed if you want more than one copy.
8.   Select small or large to determine the size of the print out.
9.   Select **OK** to print your screen image.

(Use the HELP option for more information).

4.2 LOAD A TI-92 PROGRAM FROM THE COMPUTER
1. Boot up the Link-92 program on your computer.
2. Attach the PC link cable to your calculator.
3. Select **LINK** and then **SEND** from your computer screen menu.
4. When the Send Files box is displayed, select the appropriate drive and directory.
5. You should see a list of TI-92 files on your computer screen. Select a file and then select **ADD**. If you want to send more than one file, select the file and then select **ADD** again.
6. When you have selected all of the files you want, choose **OK**.
7. The data files will be stored in your calculator unless you are sending a file whose name matches a file which is already stored in your calculator. In this case, you will be given the option to Overwrite, Rename, or Cancel.
8. When you have completed this process, select **[OK]** to return to the TI-Graph Link menu.

4.3 RUN A PROGRAM
1. The program must be stored in your calculator. (See Appendix 4.2).
2. Press **[APPS] [1]** to go to the home screen and **[CLEAR]** to clear the command line.
3. Press **[2nd]] [-]** to display a list of variables and programs stored in your calculator.
4. Press **[F2]** and when Folder appears, move the cursor to the right and then to the FOLDER named **insight** and press **[ENTER]**.
5. Move the cursor down to the Var Type and to the right to choose Program by pressing **[6]**.
6. Press **[ENTER]** to list the names of programs stored in FOLDER **insight**.
7. Move the cursor down to the name of the program that you want to run. If there are more than nine programs in your folder, some will not appear on your initial screen. In this case, continue moving the cursor down to view the the rest of the names.
8. When the cursor is highlighting the name of the program that you want press **[ENTER]**. You should see insight/program name(on the edit line.
9. Press **[)] [ENTER]** to run the program.
10. If you get an error message while running a program, press **[ESC]** to exit.
11. You can stop the execution of a program by pressing **[ON]**. If you do so, press **[ESC]** to exit.

5.1 TABLES
The TABLE option is usually used to display the values of functions which are selected for graphing. It may also be used to compare function values of two or more functions.
Walk-through example: Evaluate $f(x)=3x^2-4$ for $x=5, 5.1, 5.2,\ldots$.
1. Store this function at y1= on the y= Editor for graphing (see Appendix 1.1).
2. Press **[F2]** and press **[6]** to display the graph with a standard window.
3. Press **[APPS] [5]** to display the Table screen. Next press **[F2]** for TABLE SETUP. If this box looks "fuzzy", the cursor is probably blinking on ASK. If so, move the cursor to the right, press **[1]**, and move the cursor up to

tblStart.

4.   The value of tblStart is the initial choice of the
      independent variable for evaluating one or more functions.
      For this example, type **[5]** and move down to Δtbl.

5.   Type **[.] [1]** and move the cursor down to the next line,
      move the cursor to the right, highlight OFF and press
      [ENTER].

6.   Press **[ENTER]** to save these changes.   The table screen
      should be displayed.

7.   In the column headed by x, you should see 5, 5.1, 5.2 etc.

8.   The first three numbers in the column headed by y1 should
      be 71, 74.03 and 77.12.   Move the cursor down to see more
      data points.   If there is more than one function turned on
      you will see a third column of numbers.

9.   Press **[F2]** to display the TABLE SETUP.   Move the cursor
      down to the Graph <-> Table: line and move the cursor to
      the right.   Highlight ON and press **[ENTER]** and **[ENTER]**
      again to display the Table screen.   You should see the same
      values of x and y that you would get if you traced this
      function.

10.   Press **[F2]** to display the TABLE SETUP.   Move the cursor
      down to the Independent line, move the cursor to the right,
      highlight ASK and press **[ENTER]**.   Press **[ENTER]** again to
      return to the Table screen.

11.   Place the cursor in the column headed by x and type in **[1]**
      [0] [ENTER].   You should find the corresponding y value is
      296.   You can continue to choose the x-values (independent
      variable) and display the corresponding y-values.

12.   If you have several functions selected for graphing, their
      y-values will appear in the column headed by the name of
      the function.   For example, you may have columns for y1,
      y2, and y3,....

6.1.1 STORE DATA IN A LIST
Walk-through example: to store {.8, 1, 1.2, 1.5, 1.9} in list ℓ1:

1.   Press **[APPS] [1]** to display the home screen and press
      [CLEAR] one or two times to clear the entry line.

2.   Enter the following keystrokes to create and store the
      above set of numbers in list L1: **[2ND] [(] [.] [8] [,] [1]**
      [,] [1] [.] [2] [,] [1] [.] [5] [,] [1] [.] [9] [2ND] [)]
      [STO▸] [L] [1] [ENTER].

3.   You should see {.8 1 1.2 1.5 1.9}→ℓ1
                         {.8 1 1.2 1.5 1.9}
      To make changes in a list the command must be on the entry
      line.

4.   If the command is on the entry line, use the cursor to move
      to any character on the line.   Use **[◊] [←]** to delete a
      character.   Use **[2ND] [←]** to switch between the insert and
      type over mode.   A thin cursor indicates you are in the
      insert mode.   A thick blinking cursor indicates that you
      are in the type over mode.

5.   If the command is not on the entry line, clear the entry
      line and move the cursor to the command on the Home screen
      that you want to edit and press **[ENTER]**.   You can now move
      the cursor and use the appropriate keys to make changes.

6.1.2 CLEAR THE CONTENTS OF A LIST
To clear the contents of a list, say list ℓ1:

      From the home screen press **[2ND] [-]**, move the cursor down
      to ℓ1, and press **[F1] [1] [ENTER]** to delete the contents of
      list ℓ1.

6.2 THE ALGEBRA OF LISTS

Walk-through example: Store the numbers 1, 2, 3, 4, 5 in list ℓ1 and create new lists by applying algebraic operations to ℓ1.

1. Use Appendix 6.1.1 as a guide to store the above numbers in list ℓ1.

2. Press **[APPS] [1]** to return to the Home screen and clear the entry line by pressing **[CLEAR]**.

3. To double each number in list ℓ1 and store the results in list ℓ2, enter the following keystrokes: **[(] [2] [L] [1] [)] [STO▸] [L] [2] [ENTER]**. You should see 2·11→12 {2 4 6 8 10} on the last line of the Home screen.

4. The algebra of lists can be extended to more complicated expressions such as $\ln(5x^2+4x+7)$. To apply this function to the numbers in list ℓ1 and store the results in ℓ3, type the following keystrokes: **[CLEAR] [ln] [5] [(] [L] [1] [)] [∧] [2] [+] [4] [L] [1] [+] [7] [)] [STO▸] [L] [3] [ENTER]**.

5. On the Home screen you should see: {4ln(2) ln (35) 6ln(2) ln(103) ln(19) + 3ln(2)}. (To see all of this line move the cursor up to highlight the line and then move the cursor to the right.)

6. To evaluate the logarithm functions and approximate each of the 5 numbers: Move the cursor down to the command line and press **[CLEAR] [F2] [5] [L] [3] [)] [ENTER]**. You should see approximations of the numbers in ℓ3, the first one is 2.77259 and the last one is 5.02388.

7 - Curve Fitting (Regression) A set of data points must be stored in two lists of equal length before you can use the curve fitting features of your calculator.

Walk-through example: Fit a curve to the data points stored in list L1={.8, 1, 1.2, 1.5, 1.9} (independent variable) and L2={.4, 1.1, 1.4, 2.3, 4} (dependent variable).

7.1.1 STORE DATA PAIRS IN TWO LISTS

1. Use Appendix 6.1.1 as a guide for storing the above sets of numbers in L1 and L2.

7.1.2 STORE LISTS IN A DATA VARIABLE

1. To give the lists in Appendix 7.1.1 a data variable name of "temp" and store their contents in temp, press **[APPS] [1]**, and enter the following keystrokes at the home screen: **[CLEAR] [N] [E] [W] [D] [A] [T] [A] [SPACE BAR] [T] [E] [M] [P] [,] [L] [1] [,] [L] [2] [ENTER]**.

2. Press **[APPS] [6] [ENTER]**. This screen is called the Data/Matrix Editor and you should see the elements of L1 in column 1 and those of L2 in column 2.

3. In the Data/Matrix Editor L1 is the first column of temp and is called temp[1]. L2 is the second column of temp and is called temp[2]. **CAUTION**: When you use the Data/Matrix Editor to make changes in the first column of temp, the list L1 is not updated. Furthermore, when you make changes in list L1 from the Home screen, the first column of temp is not updated unless you use the New Data command.

7.2 LINEAR REGRESSION

1. At this point you should have your lists stored in a data variable and be in the Data/Matrix Editor. If you are not there, follow the steps in Appendix 7.1.2.

2. Press **[F5]** to display the calculate Dialog Box. Your cursor should be on the calculation type line. Move the cursor to the right to display the menu of the available

calculations.
3.   Press **[5]** (LinReg).
4.   Move the cursor down to the x line. You may now enter L1 or C1 (first column). If you have changed L1 and want to use the changes, you should enter L1. If you have used the Data/Matrix Editor to change C1 and want to use these changes, you should enter C1.
5.   Move the cursor down to the y line. You may enter L2 or C2 (second column). Reread step 4 to help you decide which name to use.
6.   Move the cursor down to the Store RegEQ line, move the cursor to the right to display the function storage names from the y= Editor. Move the cursor down to y(1)(x) and press **[ENTER]**. This will store the regression equation at y1. If you already have a function stored at y1, you may use y2 or y3, etc.
7.   Move the cursor down to the Use Freq and Categories line, move the cursor to the right, put the cursor on 1:NO and press **[ENTER]**.
8.   Press **[ENTER]** again to save the changes in the dialog box.
9.   The STAT VARS box should look like:

$$y=ax+b$$

a	$=3.173797$
b	$=-2.22246$
corr	$=.989715$
R^2	$=.979536$

The slope of this line is a and b is the y-intercept of the line. Corr is the correlation coefficient and is a measure of how well the line fits the data. R^2 is the coefficient of determination which is the square of the correlation coefficient.

7.3 QUADRATIC REGRESSION
Follow Appendices 7.1.1 and 7.1.2 to store the lists from the walk-through example in lists L1 and L2 and store the data in a Data/Matrix variable.
1.   At this point you should be in the Data/Matrix Editor. If not, press **[APPS] [6] [ENTER]**.
2.   Press **[F5]** to display the Calculate Dialog Box. Your cursor should be on the Calculation Type line. Move the cursor to the right to display the menu of available calculations.
3.   Press **[9]** to select QuadReg.
4.   Follow steps 4-8 of Appendix 7.2.
5.   The STAT VARS box should look like:

$$y=a \cdot x^2+b \cdot x+c$$

a	$=1.157726$
b	$=.030601$
c	$=-.269184$
R^2	$=.993391$

7.4 CUBIC REGRESSION
1.   Follow all of the steps in Appendix 7.2 with the exception of step 3. When you have the Calculate Dialog Box displayed, press **[3]** to select CubicReg.
2.   After completing step 8, the STAT VARS screen should look like:

$$y=a \cdot x^3+b \cdot x^2+c \cdot x+d$$

a	$=2.45395$
b	$=-8.72626$
c	$=12.665467$

$$d \quad = -5.377959$$
$$R^2 \quad = .997653$$

7.5 QUARTIC REGRESSION (A:QuartReg)

Follow all of the steps in Appendix 7.2 with the exception of step 3. This option displays a, b, c, d, e, R^2 and $y=a \cdot x^4 + b \cdot x^3 + c \cdot x^2 + d \cdot x + e$. This is the "best" fourth degree polynomial to fit your data.

7.6 LOGARITHM REGRESSION (6:LnReg)

Follow all of the steps in Appendix 7.2 with the exception of step 3. This option displays $y=a+b \cdot \ln(x)$ and values of a and b.

7.7 EXPONENTIAL REGRESSION (4:ExpReg)

Follow all of the steps in Appendix 7.2 with the exception of step 3. This option displays $y=a \cdot b^x$ and values of a and b.

7.8 POWER REGRESSION (8:PowerReg)

Follow all of the steps in Appendix 7.2 with the exception of step 3. This option displays $y=a \cdot x^b$ and values of a and b.

7.9 OVERLAY THE GRAPH OF A REGRESSION EQUATION ON A SCATTER PLOT.

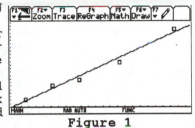

Figure 1

Walk-through example: To plot the linear regression line from Appendix 7.2 over the scatter plot for L1 and L2 in the walk-through example:

1.   Store the lists in L1 and L2 and store the data in a Data/Matrix variable (see Appendices 7.1.1 and 7.1.2).

2.   Use Appendix 7.2 as a guide for doing a linear regression analysis.

3.   When you are fitting a curve to a set of data and the Calculate Dialog Box is displayed, you have an option to store the regression equation at yn(x) for the n of your choice. If you did not do this when you performed your regression, repeat the regression, this time storing the regression equation at y1 or y2, etc.

4.   Press **[APPS] [2]** to display the y= Editor. There should be a check mark by the regression equation. If not, put the cursor on this equation and press **[F4]**.

5.   Move the cursor up so that Plot 1: is highlighted and press **[ENTER]** to display the Plot Dialog box.

6.   Move the cursor to the right to display the plot types and press **[1]** to select Scatter.

7.   Move the cursor down to the mark line and move the cursor to the right and press **[1]** to select Box.

8.   Move the cursor down to the x line and enter the list name for the independent variable (L1 or C1 for the walk-through example.)

9.   Move the cursor down to the y line and enter the list name for the dependent variable (L2 or C2 for the walk-through example).

10.   Move the cursor to the Use Freg and Categories, move it to the right, and press **[1]** to select NO.

11.   Press **[ENTER]** to return to the y= Editor. There should be a check mark by Plot 1. If not, put the cursor on Plot 1 and press **[F4]**. (See Appendices 1.5 and 1.6 for more information about scatter plots).

12.   Press **[F2]** and then **[9]** to set an appropriate window for the graph.

TI-92

13.   Your graph should look like the one in Figure 1.

8.1 NDERIV
Walk-through example: Approximate $f'(2)$ where $f(x)=3x^2+x$.
1.   Press **[APPS] [1]** to display the Home screen.
2.   Press **[2nd] [5]** to display the MATH menu. Press **[A]** to display the CALCULUS submenu.
3.   Press **[A]** to choose nDeriv(
4.   The syntax for the nDeriv command is: nDeriv($f(x),x$)|x=a. For example, nDeriv($3x^2+x,x$)|x=2 returns an approximation to $f'(x)$ at x=2.
5.   With "nDeriv(" displayed enter the following keystrokes: **[3] [x] [∧] [2] [+] [x] [,] [x] [)] [2nd] [K] [x] [=] [2]** and press **[ENTER]**.
6.   You should see nDeriv($3 \cdot x^2+x,x$)|x=2   13.

8.2 $\dfrac{dy}{dx}$ (Numerical Derivative) FROM THE GRAPH OF y=f(x)

Walk-through example: Approximate the slope of the tangent line for $y=3x^2+x$ at x=1.
1.   Graph $y=3x^2+x$ with a standard window.
2.   Press **[F5]** to display the MATH menu.
3.   Press **[6]** to display the Derivatives menu and press **[1]** to select dy/dx.
4.   The graph of $y=3x^2+x$ should be displayed and the prompt "dy/dx at?" should be in the lower left corner of the screen. Move the cursor to the point where xc: is 1.092437 and press **[ENTER]**.
5.   You should see dy/dx=7.5546218.

8.3 $\left(\dfrac{dr}{d\theta}\right)$

Walk-through example: Approximate $\dfrac{dr}{d\theta}$ at $\dfrac{\pi}{6} \approx .523$ for $r=4\sin(2\theta)$.

1.   Follow the instructions in Appendix 1.4 to graph $r=4\sin(2\theta)$ with the specified window settings.
2.   With the graph of $r=4\sin(2\theta)$ displayed, press **[F5] [6] [4] [ENTER]** to select $\dfrac{dr}{d\theta}$ from the derivatives sub-menu.
3.   Move the cursor to the point whose θc: is about .524 and press **[ENTER]**.
4.   You should see dr/dθ=4

8.4 nInt (Numerical Integral)
Walk-through example: Find $\int_{1}^{3} x^3 dx$

1.   Press **[APPS] [1]** to display the Home screen and press **[CLEAR]** to clear the entry line.
2.   Press **[2nd] [5]** to display the MATH menu and press **[A]** to display the Calculus sub menu.
3.   Press **[B]** to choose nInt(.
4.   The syntax for nInt is nInt($f(x),x,a,b$) where the entries correspond to the definite integral $\int_{a}^{b} f(x)\ dx$.

5.   Consider the definite integral $\int_{1}^{3} x^3\ dx$.   With nInt(

displayed enter the following keystrokes: **[x]** **[∧]** **[3]** **[,]** **[x]** **[,]** **[1]** **[,]** **[3]** **[)]** **[ENTER]**.
6.   You should see nInt(x^3,x,1,3)   20.

8.5 ∫f(x)dx (Numerical Integration) FROM THE GRAPH OF y=f(x)

Walk-through example: Find $\int_{1}^{3} x^3\, dx$.

1.   Store y=x^3 on the y= menu and highlight this function for graphing.
2.   Set the window variables as:
          xmin=-11.8   ymin=-5   xres=2
          xmax= 12   ymax=40
          xscl=1   yscl=2
3.   Press **[APPS]** **[4]** to display the graph and press **[F5]** to display the GRAPH/MATH menu.
4.   Press **[7]** to select ∫f(x)dx. Move the cursor to the point whose xc is 1 and press **[ENTER]**.
5.   Move the cursor to the point whose xc is 3 and press **[ENTER]**.

6.   You should see ∫f(x)dx=20 and the region defined by $\int_{1}^{3} x^3\, dx$

shaded in.

8.6 DRAW A TANGENT LINE TO A POLAR GRAPH
Walk-through example: Draw a line tangent to the polar graph of r=4sin(2θ) at the point where θ=$\frac{\pi}{6}$≈.524.

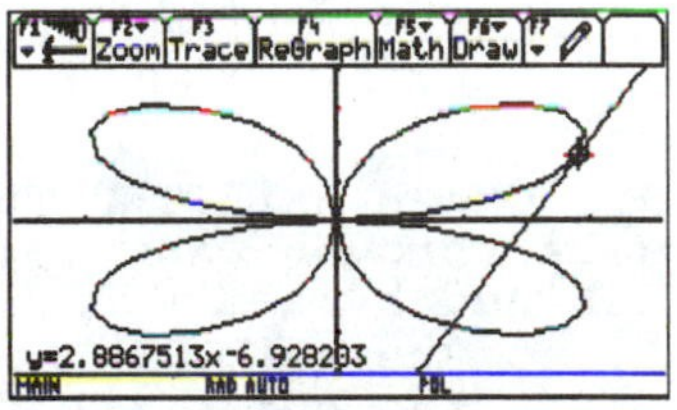

1.   Follow the instructions in Appendix 1.4 to graph the above function.
2.   Press **[F1]** **[9]** to display the GRAPH FORMATS dialog box.
3.   On the coordinates line, move the cursor to the right and press **[2]** **[ENTER]** to select POLAR which will display polar coordinates of points in the trace mode.

Figure 1

4.   Press **[F5]** **[A]** to select Tangent. Move the cursor along the curve to a point where the θ coordinate (θc) is about .524 and press **[ENTER]**.
5.   Your graph should look like the one in Figure 1.

9.1 THE SEQUENCE COMMAND
The seq(command is used to evaluate selected terms of a sequence or store selected terms of a sequence in a list.
Walk-through example: Display the 3rd, 5th, 7th, 9th and 11th terms of the sequence a_n=n^3-2.
1.   Press **[APPS]** **[1]** to display the home screen and **[CLEAR]** to clear the entry line.
2.   Press **[2nd]** **[5]** to display the MATH menu and press **[3]** to select the List submenu.
3.   Press **[1]** to display the seq(command.
4.   With seq(displayed on the entry line enter the following keystrokes: **[N]** **[∧]** **[3]** **[-]** **[2]** **[,]** **[N]** **[,]** **[3]** **[,]** **[1]** **[1]** **[,]** **[2]** **[)]** **[ENTER]**
5.   You should see:
        seq(n^3-2,n,3,11,2)   {25 123 341 727 1329}
        a_n=n^3-2 defines the sequence, n represents the variable, 3 represents the first term to be evaluated, 11 the last term

to be evaluated and 2 is the increment. The sequence command with these parameters displays a_3, a_5, a_7, a_9, and a_{11}, which have the values shown above.

9.2 THE SUM COMMAND

Walk-through example: Calculate $\displaystyle\sum_{n=1}^{5}\frac{1}{n}=1+\frac{1}{2}+\frac{1}{3}+\frac{1}{4}+\frac{1}{5}$.

1.   Press **[APPS] [1]** to display the home screen and **[CLEAR]** to clear the entry line.
2.   Press **[2nd] [5]** to display the MATH menu and press **[3]** to select the List submenu.
3.   Press **[6]** to select sum.
4.   Press **[2nd] [5] [3]** to display the MATH/LIST menu and press **[1]** to select seq(.
5.   With sum(seq(displayed on the entry line enter the following keystrokes: **[1] [÷] [N] [,] [N] [,] [1] [,] [5] [,] [1] [)] [)]** **[ENTER]**.
6.   You should see:

$$\text{sum}\left(seq\left(\frac{1}{n},n,1,5,1\right)\right) \qquad \frac{137}{60} \text{ or } 2.28333$$

$$\text{(depending on the mode)}$$

7.   To change $\dfrac{137}{60}$ to a decimal press **[CLEAR]** to clear the entry line.
8.   Press **[F2] [5]** to display "approx(".
9.   Move the cursor up to $\dfrac{137}{60}$ and press **[ENTER]**.
10. Press **[)] [ENTER]** to see:

$$\text{approx}\left(\frac{137}{60}\right) \qquad\qquad 2.28333$$

9.3 GRAPH SELECTED TERMS OF A SEQUENCE

Walk-through example: Graph the first 20 terms of the sequence defined by $a_n=2n-5$.

1.   Press **[MODE]** and with the cursor on the Graph... line, move the cursor to the right to display a menu of graph types.
2.   Press **[4]** to select the SEQUENCE graphing mode and press **[ENTER]** to save this change.
3.   Press **[APPS] [2]** to display the Y= Editor.
4.   With the cursor at u1= enter the keystrokes **[2] [N] [-] [5]** **[ENTER]** to display 2·n-5 at u1=.
5.   Press **[F6]** and **[3]** to plot a square at each point and do not connect the point.
6.   A check mark should be to the left of u1. If not, place the cursor on the u1= line and press **[F4]**.
7.   Press **[APPS] [3]** to display the window variables for sequence graphing.
8.   Enter the values of the window variables as follows:

nmin=1	xmin=1	ymin=-4
nmax=20	xmax=20	ymax=40
plotstrt=1	xscl=1	yscl=2
plotstep=1		

9.   Press **[APPS] [4]** to display the graph of the first 20 terms of this sequence.
10. Press **[F3]**

Graph the first 20 terms of the sequence defined by $a_1=2$ and
$a_n=a_{n-1}+\dfrac{1}{n}$.

1.   Press **[MODE]**. With the cursor on the Graph... line move
     the cursor to the right to display a menu of graph types.
     Press **[4]** to select the SEQUENCE graphing mode and press
     [ENTER] to save this change.
2.   Press **[APPS] [2]** to display the -y= menu. With the cursor
     on u1= press **[CLEAR]** and enter the keystrokes: **[U] [1] [(]**
     [N] [-] [1] [)] [+] [1] [÷] [N] [ENTER].
3.   At u1= you should see $u1(n-1)+\dfrac{1}{n}$.
4.   At ui1= enter the value of the first term, which is 2, and
     press **[ENTER]**.
5.   Press **[F6]** and **[3]** to select the graph style.
6.   Press **[APPS] [3]** to display the window variables for
     sequence graphing and enter the following settings:
              nmin=1   xmin=1   ymin=-1
              nmax=20   xmax=20   ymax=5
              plotstrt=1   xscl=1   yscl=1
              plotstep=1
7.   Press **[APPS] [4]** to display the graph of the first 20 terms
     of this sequence.
8.   Press **[F3]** (trace) to evaluate the terms.
9.   Compare your results with: $u_1=2$, $u_6=3.45$ and $u_{15}=4.31823$.

10.1 STORE A PICTURE
Walk-through example: Store the graph of $y=x^2$ as a picture.
1.   Store $y=x^2$ at y1=, set a standard window (Zoom 6) to
     display the graph of $y=x^2$ (See Appendix 1.1).
2.   Press **[F1] [2]** to display the SAVE COPY AS box.
3.   With the cursor on the Type: line, move the cursor to the
     right and press **[2]** for Picture.
4.   Move the cursor down to the variable box and name the
     picture by pressing **[P] [I] [C] [1] [ENTER]**.
5.   Press **[ENTER]** to save the picture.
6.   If there is another picture with the same name, it will be
     replaced by this one.

10.2 RECALL A PICTURE
Walk-through example: Recall the picture that was stored in
Appendix 10.1.
1.   Press **[APPS] [2]** to display the Y= editor.
2.   Press **[F5] [1]** to turn off all functions and scatter plots.
3.   Turn the axes off by pressing **[APPS] [4]** and choosing **[F1]**
     [9] to view the graphing format. Move the cursor down to
     Axes, right to display the drop-down box, and press **[1]**
     [ENTER] to turn the axes off.
4.   Press **[APPS] [4]** to display the graph screen which should
     be clear. If the graph screen is not clear, press **[F6] [1]**
     to clear any previous drawings.
5.   Press **[F1] [1]** to display the OPEN dialog box. With the
     cursor on the Type: line, move it to the right and press
     [2] to select Picture.
6.   Move the cursor down to the Variable: line and to the right
     to display a current list of picture names. Put the cursor
     on pic1 and press **[ENTER]**. Press **[ENTER]** again and the
     graph of $y=x^2$ should appear on the screen.
7.   After you finish viewing the picture, follow step 3 to turn
     the axes back on. This will also clear the picture.

TI-92

10.3 OVERLAY A PICTURE ON A GRAPH

Walk-through example: overlay the picture which was stored in Appendix 10.1 on the graph of $y=(.3)^x$.
1.   Store the graph of $y=x^2$ with a standard window as a picture with name pic1 (see Appendix 10.1).

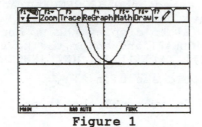

Figure 1

2.   Store $y=(.3)^x$ by using the y= editor. Press **[F2] [6]** to display the graph and **[F6] [1]** to clear any drawings.
3.   Recall the picture named as pic1 by following steps 5 and 6 of Appendix 10.2.
4.   Your graph should look like the one in Figure 1.
5.   The window settings for the current graph and the picture must be identical in order for the combined graph to have any meaning.

11.1 APPROXIMATE A SOLUTION TO AN EQUATION

Walk-through example: To solve $\cos x = x-1$, $-10 \le x \le 10$,:
1.   Write the equation as $\cos x - x + 1 = 0$.
2.   Press **[APPS] [2]** to display the y= edit screen.
3.   Store $y = \cos x - x + 1$ at a y= position. Turn off all other functions and plots.
4.   Press **[F2] [6]** to set up a standard graphing window and display the graph.
5.   Press **[F5] [2]** to select zero.
6.   You should see the prompt "Lower Bound?". Position the cursor on a point just to the left of the intercept and press **[ENTER]**.
7.   You should now see the prompt "Upper Bound?". Move the cursor to a point just to the right of the intercept and press **[ENTER]**. The solution x:1.2834287 should be displayed on the screen. (You can select the number of digits by selecting Display Digits from the MODE menu.)

11.2 APPROXIMATE A SOLUTION TO A SYSTEM OF TWO EQUATIONS

Walk-through example: Solve $y = \cos x$ and $y = x-1$ for $-10 \le x \le 10$.
1.   Write the equation $\cos x = x-1$ and follow Appendix 11.1.

TI-92